The Electronic Scholar

Enhancing Research Productivity with Technology

Dave L. Edyburn
University of Wisconsin-Milwaukee

Merrill,
an imprint of Prentice Hall
Upper Saddle River, New Jersey   Columbus, Ohio

Library of Congress Cataloging-in-Publication Data

Edyburn, Dave L.
      The electronic scholar : enhancing research productivity
   with technology / Dave L. Edyburn.
          p.   cm.
   Includes bibliographic references and index.
   ISBN 0-13-080601-3
      1. Research--methodology.   2. Research--Data processing.
3. Computer science.   I. Title.
Q180.55.M4E25 1999
001.4'2--dc21   98-38032
                                      CIP

Cover art: © Artville
Editor: Kevin M. Davis
Production Editor: Julie Peters
Cover Design Coordinator: Diane C. Lorenzo
Text Designer: Dave Edyburn
Cover Designer: Dan Eckel
Production Manager: Laura Messerly
Director of Marketing: Kevin Flanagan
Marketing Manager: Suzanne Stanton
Advertising/Marketing Coordinator: Krista Groshong

This book was set in Palatino by the author and was printed and bound by R.R. Donnelley &
Sons Company. The cover was printed by Phoenix Color Corp.

 © 1999 by Prentice-Hall, Inc.
Simon & Schuster / A Viacom Company
Upper Saddle River, New Jersey 07458

Printed in the United States of America

10 9 8 7 6 5 4 3 2 1

ISBN: 0-13-080601-3

Prentice-Hall International (UK) Limited, *London*
Prentice-Hall of Australia Pty. Limited, *Sydney*
Prentice-Hall of Canada, Inc., *Toronto*
Prentice-Hall of Hisanoamericana, S. A., *Mexico*
Prentice-Hall of India Private Limited, *New Delhi*
Prentice-hall of Japan, Inc., *Tokyo*
Simon and Schuster Asia Pte. Ltd., *Singapore*
Editora Prentice-Hall do Brasil, Ltda., *Rio de Janeiro*

Preface

Have you visited one of the technology superstores lately? They are as large as grocery stores, lined with rows and rows of computers, monitors, printers, disk drives, and scanners.

Need some accessories for your computer? Be prepared to navigate a row of printer cartridges, a wall of cables, and shelves full of disks.

Just stopped by to pick up some new software? You'll find hundreds of choices in rows conveniently organized by computer platform (Macintosh, Windows) and application (productivity, educational, home, games, etc.).

Before you leave, don't forget to browse the mini-bookstore, which offers a wide variety of magazines and books, general and specific, all designed to make you a more confident and productive user of technology.

The experience of visiting a technology superstore is both exciting and overwhelming: exciting because it expands one's awareness of the possibilities technology offers, and overwhelming because there are so many choices.

Some members of the research community have eagerly sought out the technology superstore, found ways to navigate the plethora of choices the marketplace offers, and acquired specific products to use in their research. Others have found the experience a bewildering trek into uncharted territory.

Is technology essential for our work or merely optional for those folks who like playing with gadgets? The use of technology for data analysis is a historical tradition within the research community. For the past twenty years various statistical analysis packages have represented the single tool in the researcher's toolkit. Recently, three other tools have emerged: a word processor, an email program, and a web browser. These tools represent a significant milestone: the research community has begun to recognize the value of an integrated set of electronic tools. And mastery of these tools is growing ever more essential to the work of the researcher.

Technology-Enhanced Research Productivity

This book is designed to serve as a concise introduction to the technological possibilities associated with conducting and administering research. The identification and use of appropriate electronic tools can profoundly impact a profession. My aspiration is to transform technology-using researchers into electronic scholars. I describe three goals to guide our efforts: (1) advance a vision of how information technology is used as an essential tool by researchers in their work, (2) extend our understanding of the productivity demands made on researchers and of the potential tools that facilitate productivity, and (3) rethink how researchers work in light of available information technologies.

This book is ideally suited to serve as a supplementary text for any research course; however, it is not an introduction to research methodology. Also, it assumes previous experience with the core tools of electronic scholarship: word processor, email, web browser, and data analysis products. Readers who are comfortable with basic desktop navigation and a web browser will find this book to be an invaluable resource. Experienced researchers will find the format logical and the resources valuable. Researchers with minimal technology experience are advised to seek the assistance of a mentor to acquire the skills and knowledge outlined in Table 1-7 in Chapter 1.

Organization of This Book

The goal of this book is to help you create an electronic toolkit. To that end, we first describe a process for building a general productivity toolkit to support your work. Next we will explore the technological possibilities for building a research productivity toolkit. You will note that this book is organized around 28 tasks commonly completed by researchers. The book could have been organized by tools, for example, a section on word processing, then one on web browsing; however, using a taxonomy based on what researchers do should enable you to quickly identify the sections that are most relevant to your research. Within each task, tactics or strategies for using technology to enhance your research productivity are discussed.

Chapter 1 introduces the concept of the electronic scholar as a means of rethinking the use of technology in research. This chapter provides a framework for assembling an electronic toolkit for enhancing general productivity and presents the 28 tasks commonly associated with research productivity.

Chapter 2 focuses on using technology to foster professional development. It describes how to enhance skills and knowledge to stay current in the field, and provides tactics that include specific examples of technology tools for ongoing professional development.

The core tasks of the research process are presented in Chapters 3 through 6 with each phase having its own chapter: designing the study in Chapter 3, conducting the study in Chapter 4, analyzing the data in Chapter 5, and reporting the results in Chapter 6. Within this framework, 16 common tasks are identified and 46 tactics for enhancing research productivity are described.

Chapter 7 examines the use of technology to facilitate the administrative aspects of directing a research project. Tasks involved in the administration of both large and small research projects are described.

Finally, Chapter 8 provides practical advice about the quest of becoming an electronic scholar. Since the resources outlined in this book are only a starting point, this chapter describes the effort that will be required to integrate the tools and productivity strategies into the daily routine of the researcher.

A Single Focus: Technology-Enhanced Research Productivity

Much like a travel guide, this book provides clear signposts to places you might want to visit. We will explore web sites with extraordinary information resources and visit companies that produce products or services that offer intriguing possibilities for your research. When available, I'll point out free software that you can download and try out. Throughout our explorations we'll discover some of the creative possibilities that result when talented people understand how to apply technology to particular aspects of the research enterprise. The journey will surely include some "gee-whiz" technology, but do not be misled. The single focus of this book is to provide the research community with a source book of possibilities for using technology to enhance research productivity. Simply put, this book is about the journey of integrating technology into one's daily work. The key to transforming technology-using researchers into electronic scholars is capturing the potential of technology to enhance research productivity.

Some readers may want to systematically work their way through the book as I offer a guided exploration of more than 750 web sites. Other readers may prefer to become familiar with the taxonomy of productivity tasks and then access specific tactics and tools. This book is about possibilities. Experiment with the possibilities using the many examples I have provided for you. Decide which technology tools will support your research productivity.

I am delighted to share this book with you. I encourage you to share your insight with others about the technology tools you have discovered and the ways they enhance your productivity as a researcher. I look forward to the opportunity to discuss the usefulness of this approach for understanding the impact of technology on the research process. Finally, the question, "How does technology enhance research productivity?" still remains one that must be answered personally. As the research community attempts to define new standards for electronic scholarship, I invite you to add your voice to the conversation.

Acknowledgments

This book could not be completed without the inspiration and assistance of many people. Several colleagues recognized the potential value of this work at a very early stage and provided significant encouragement: Cindy Okolo, University of Delaware; Kyle Higgins, University of Nevada-Las Vegas; Randy Boone, University of Nevada-Las Vegas; and John Lloyd, University of Virginia.

The thoughtful comments of several reviewers sharpened the focus of the book: James H. Banning, Colorado State University; Sylvia A. Gholson, New York University; Jeri Levesque, Webster University; Malino Monaco, Georgia State University; Charles L. Thomas, George Mason University; and Paul Westmeyer, University of Texas, San Antonio.

I owe a special debt of graditude to Gary Showers and Lisa Dieker who provided many valuable insights throughout the development process. In addition, a special thank you to two "electronic scholars in-training" who spent innumerable hours reviewing drafts of the manuscript and offering their views on the usefulness of the sites: Simone Conceição-Runlee and Allen Piepenbrink.

This project could not have been completed without the valuable assistance of the editorial team at Merrill/Prentice Hall. Editor Kevin Davis offered insight into the potential of this work and encouragement during the entire process. Julie Peters and Holly Jennings answered countless questions and provided timely advice. It has been a pleasure to work with each of you.

To Pam, Keith, and Kelly: thank you for the time and inspiration to complete this book. As I look forward to the brief respite associated with being the "electronic scholar unplugged," I eagerly await our new adventures together.

D.L.E.
August, 1998

A Commitment to Accuracy

The web site addresses in this book were accurate as of press time. However, due to rapid change associated with the Internet, web site addresses can change without notice. To ensure that readers have accurate information, a companion web site has been established to provide periodic updates to the information in this book:

http://merrilleducation.com/Edyburn

Readers interested in offering feedback or sharing resources for future editions of this book may contact the author by email:

edyburn@uwm.edu

Contents

Chapter 1 The Electronic Scholar 1

Tools of the Trade 1
Perspectives on Technology-Enhanced Productivity 2
The Electronic Scholar 3
Next Steps 18
References 20

Chapter 2 Professional Development 21

1. Enhancing Knowledge and Skills 22
1-1 Identifying resources for learning about research 22
1-2 Identifying resources for teaching about research 23
1-3 Exploring research methods and procedures 26
1-4 Participating in the research community 33
1-5 Considering career opportunities 35
1-6 Learning about anything you can imagine 37
1-7 Developing a personal library 43

2. Maintaining Current Awareness 45
2-1 Subscribing to electronic information services 45
2-2 Utilizing news services 47
2-3 Utilizing tools that monitor 51
2-4 Shifting from online to offline browsing 52

Chapter 3 Designing the Study 53

3. Engaging in Preliminary Exploration of a Topic 54
3-1 Sending email 54
3-2 Browsing the web 55

4. Conducting an Extensive Review of the Literature 61
4-1 Identifying appropriate databases to search 61
4-2 Browsing web-based publications 62
4-3 Considering options for document delivery 64
4-4 Writing the literature review 68

5. Selecting an Appropriate Methodology 68
5-1 Selecting a research design 68
5-2 Locating resources for qualitative research 69
5-3 Locating resources for quantitative research 72

5-4 Locating resources for survey research 73
5-5 Selecting and describing subjects 76
5-6 Locating assessment instruments 78

6. Identifying Potential Sources of Funding 79
6-1 Subscribing to a funding alert service 79
6-2 Searching for appropriate funding sources 80

7. Engaging in Collaboration 82
7-1 Locating an individual...82
7-2 Using the web as a telephone 83
7-3 Using video conferencing...84
7-4 Using audiographics 86
7-5 Using work group software 87

8. Preparing Proposals...88
8-1 Cutting and pasting information 88
8-2 Using proposal development tools 89

9. Participating in Human Subjects Review 89
9-1 Locating procedures, policies, and forms 89

Chapter 4 Conducting the Study 91

10. Creating Materials 92
10-1 Creating print materials 92
10-2 Using rapid prototype tools 93
10-3 Creating web-based materials 95

11. Collecting Data...96
11-1 Using hand-held data input systems 96
11-2 Using portable keyboards 98
11-3 Using computer-based data collection tools 99
11-4 Using unobtrusive computer-based data collection tools 102
11-5 Using web-based data collection tools 102
11-6 Using voice input...103
11-7 Using digital cameras 104

12. Communicating with Field Sites 105
12-1 Using email...105

Chapter 5 Analyzing the Data 107

13. Analyzing Quantitative Data 108
13-1 Using generic tools for data analysis 108
13-2 Using specialized data analysis tools 109

14. Analyzing Qualitatitive Data 112
14-1 Using generic tools for text analysis 112
14-2 Using specialized text analysis tools 113

Chapter 6 Reporting the Results 117

15. Consulting Writing Resources 118
15-1 Using ready reference tools for writers 118
15-2 Using guides to grammar, word use, and style 121

16. Exploring the Visual Representation of Data 124
16-1 Using generic tools for visual representation of data 124
16-2 Using creative tools for visual representation of data 125
16-3 Using specialized tools for visual representation of data............128

17. Disseminating Research Findings 129
17-1 Disseminating information in print formats 129
17-2 Disseminating information via the web 131
17-3 Disseminating information in electronic formats 137

18. Presenting Research Results 140
18-1 Using presentation software 140

Chapter 7 Research Administrator 141

19. Managing Information 142
19-1 Creating electronic text 142
19-2 Creating databases 144

20. Managing Projects 144
20-1 Using project management software 144
20-2 Using time management software 145
20-3 Meeting technical requirements 146

21. Maintaining Financial Records 148
21-1 Using financial recordkeeping software 148

22. Selecting Mailing Systems 148
22-1 Mailing and tracking systems 148

23. Purchasing Supplies 149
23-1 Purchasing office supplies 149

24. Purchasing Equipment 150
24-1 Purchasing new equipment 150
24-2 Purchasing upgrades 151
24-3 Purchasing used equipment 152

25. Obtaining Technical Support **152**
25-1 Obtaining phone support 152
25-2 Using troubleshooting tools 153
25-3 Accessing web sites that offer technical support 154

26. Facilitating Travel Planning **156**
26-1 Making reservations 156
26-2 Exploring a city 159
26-3 Checking on the weather 160
26-4 Working while you travel 161

27. Preparing Disaster Recovery Plans **162**
27-1 Using backup systems 162
27-2 Planning for disaster recovery 163

28. Participating in Technology Transfer Processes **164**
28-1 Facilitating technology transfer 164

Chapter 8 The Quest to Become an Electronic Scholar 165

Adoption of Innovations 170
The Technology Integration Process 172
Looking Ahead 174
References 176

Index 177

Tables and Figures

Tables

Table 1-1 Contrasting Perspectives on How Technology
        Enhances Productivity 3
Table 1-2 Goals and Strategies for Fostering a Common Vision
        of Technology-Enhanced Research Productivity 4
Table 1-3 Common Tasks Involved in General Productivity 6
Table 1-4 28 Tasks Commonly Associated with
        Research Productivity 7
Table 1-5 Common Technologies for Enhancing
        Instructional Productivity 9
Table 1-6 Edyburn's Taxonomy of Technology-Enhanced
        Research Productivity 10
Table 1-7 World Wide Web Self-Assessment Skill Checklist 19

Figures

Figure 1-1 General Productivity Worksheet 14
Figure 1-2 Research Productivity Worksheet 15

Figure 2-1 Web Page of the Office of Educational Research
        and Improvement, U.S. Department of Education 23
Figure 2-2 Linguistic Resources on the Internet 30
Figure 2-3 Home Page of the American Educational Research Association 34
Figure 2-4 The All in One Search Page 39
Figure 2-5 The Scout Report Home Page 46

Figure 3-1 A Beginner's Guide to Effective Email 54
Figure 3-2 Internet Resources for Institutional Research 57
Figure 3-3 Argus Clearinghouse 59
Figure 3-4 Home Page for UnCover 67
Figure 3-5 QualPage, Resources for Qualitative Researchers 70
Figure 3-6 Y? The National Forum on People's Differences Home Page 73
Figure 3-7 The ERIC Clearinghouse on Assessment and
        Evaluation Web Site 79

Figure 4-1 Example of a Web-Based Form for Collecting Data 95
Figure 4-2 PalmIII™ connected organizer 96
Figure 4-3 AlphaSmart 2000 98
Figure 4-4 A Computer-attachable Probe from Team Labs 100

Figure 5-1 A Web Page Devoted to Statistical Calculators 109
Figure 5-2 The SPSS Web Page 110
Figure 5-3 Home Page of Qualitative Solutions Research Page 113

Figure 6-1 Determining the Appropriate Bibliographic Format
            for Citing Electronic Information 121
Figure 6-2 Example of a PowerPoint Presentation Published
            on the Web 131
Figure 6-3 A Field Guide to Institutional Research Home Pages 133
Figure 6-4 Home Page of StorySpace™ 137

Figure 7-1 The Web Site for Support Help 155
Figure 7-2 Mobile Computing Home Page 161

Figure 8-1 Sample Toolkit for Supporting General Productivity 166
Figure 8-2 Sample Toolkit for Supporting Research Productivity 167
Figure 8-3 The Revised Technology Adoption Life Cycle 170
Figure 8-4 Stages of Concern Involved in Adoption of New Innovations 171
Figure 8-5 Edyburn's Model of the Technology Integration Process 172

CHAPTER 1
The Electronic Scholar

The widespread availability of computers and research productivity software provides researchers with new opportunities to work more effectively and efficiently than in the past. However, persistent questions such as, "What tools are available?" and "How will this product help me in my work?" must be answered before the potential of technology can be realized. The purpose of this chapter is to foster a vision of the possibilities technology affords researchers. We examine the use of technology toolkits as a strategy for transforming technology-using researchers into electronic scholars.

Tools of the Trade

In many fields today, the introduction of new technologies has fundamentally altered the basic tools with which professionals work. Consider the following examples:

- Developed around the knowledge of expert automotive mechanics, comprehensive automotive troubleshooting systems facilitate the work of automotive technicians by diagnosing problems in a car's engine and suggesting repair procedures.
- Electronic spreadsheets allow financial managers to change one figure and examine the impact of the change on all aspects of a financial plan. Such "What-if" calculations can be further enhanced by the dynamic display of the data in a graph.
- Computer-assisted design (CAD) systems have fundamentally altered the work of architects and product designers by reducing the time involved in creating original designs and eliminating the laborious effort involved in preparing modifications.
- Bar code systems enable express mail companies to track hundreds of thousands of products on a daily basis, have instant access to the whereabouts of any package in the system, and design optimal routes to ensure prompt delivery.

These examples illustrate how the identification and use of appropriate electronic tools can profoundly impact a profession. There is mounting evidence in the business community that the development of information technology tools will continue to impact the knowledge and skills needed by professionals and redefine how tasks are completed (Hammer & Champy, 1993; Kennedy, 1993;

Naisbitt, 1982). Indeed, there is a clear sense of urgency within the business community to understand the potential contributions of information technology in order to gain strategic and competitive advantages in a global economy (Peppers & Rogers, 1997; Stewart, 1997).

The educational community has also sought new technology during the past decade (Quality Education Data, 1997). Schools and universities have found themselves on a technology shopping spree, spending an increased percentage of their annual budgets on technology. Unfortunately, in contrast to the examples from the business community, the evolution of information technologies over the past 10 years has yet to significantly impact the basic toolbox with which educators and researchers work. It appears that we have been generally content with the traditional tools of the trade: paper and pencil, the chalkboard, and the textbook.

Perspectives on Technology-Enhanced Productivity

Although the potential of technology to impact the work of scholars has been recognized (Brent & Anderson, 1990; Harrison & Stephen, 1996; Kelle, 1995; Lawrence, 1984; Tesch, 1990), the everyday work of the researcher appears largely unchanged. This may be explained by the fact that traditional efforts to use technology to enhance professional productivity have focused on a single tool: the integrated software package. Software packages such as the "Works" programs (e.g., *AppleWorks, Microsoft Works, ClarisWorks*) can be powerful, useful tools, but because training often has focused on teaching the mechanics of operating the software rather than on developing personally relevant applications for enhancing productivity, their fullest potential is not realized in many cases.

We need to rethink the way we view technology-enhanced professional productivity. This perspective is built on three goals: advance a vision of how information technology should be used as an essential tool by researchers in their work, improve our understanding of the productivity demands made of researchers, and rethink how researchers work in light of available information technologies. Table 1-1 contrasts the new perspective favored here with the traditional view of enhancing productivity. As outlined in Table 1-2, each goal is accompanied by one or more strategies that must be carried out if the promise of technology-enhanced productivity is to be fulfilled.

Goal 1 addresses the need to define a vision of how information technology is viewed as essential for the work of the researcher. Presently, individuals offer their vision of technology's potential for research; indeed, this book is one such effort. Individual visions are a necessary developmental milestone. Nonetheless, we must find venues within the profession to move from individual visions to common visions that have been defined and validated. Until this happens, using technology to enhance research productivity will remain a "do-it-yourself project."

Table 1-1
Contrasting Perspectives on How Technology Enhances Productivity

Traditional View	New Perspective
Enhancing Productivity...	Enhancing Productivity...
...means learning to use a single integrated program.	...means assembling a small number of products that work together as an integrated desktop.
...means shopping for the "best" tool.	...means learning how to exploit the power of each tool.
...means learning the mechanical operation of a program.	...means developing insight into the relationship between tasks, tools, and productivity.

Goal 2 involves increasing our understanding of the tasks researchers carry out and the linkage between these tasks and tools. A framework is needed to allow us to subsequently explore how specific tools may be integrated into our work as researchers.

Finally, Goal 3 involves rethinking how researchers work in light of available technologies. Whereas we may easily accomplish the first two goals, it is only when we address this third goal, personally and collectively, that we will capture the potential of technology to enhance our work as researchers.

We will now turn our attention to putting these three goals and their corresponding strategies into operation as we examine the concept of the electronic scholar.

The Electronic Scholar

The concept of the electronic scholar was advanced by Lawrence (1984) as a means of thinking about how technology can serve to enhance professional productivity. Essentially, the term *electronic scholar* describes an individual who makes full use of existing information technologies to manage an ever-increasing amount of information. Lawrence recognized that information technology was increasingly accessible and that professionals needed to identify the electronic tools that would support all aspects of their work.

To become electronic scholars we must clarify a vision of what an electronic scholar looks like, the toolbox we use, and how our productivity is enhanced. As a researcher, an electronic scholar assembles an integrated set of technology tools (an electronic toolbox, if you will) rather than simply focusing on single applications of technology (e.g., word processing, desktop publishing, statistical analysis). As we shall see later in this chapter, the goal of creating an integrated set of technology tools is to produce a synergy among the applications on the desktop.

Table 1-2
Goals and Strategies for Fostering a Common Vision of
Technology-Enhanced Research Productivity

Goal 1: **Advance a vision of how information technology is used as an essential tool by researchers in their work**

Strategy 1.1 Appreciate the systematic application of information technology in the research process

Goal 2: **Extend our understanding of the productivity demands made on researchers and the potential tools that facilitate productivity**

Strategy 2.1 Increase our understanding of the productivity demands made on researchers

Strategy 2.2 Identify appropriate tools that contribute to the completion of tasks commonly associated with research productivity

Strategy 2.3 Create a toolbox of electronic tools

Goal 3: **Rethink how researchers work in light of available information technologies**

Strategy 3.1 Facilitate training in the use of the toolbox

Strategy 3.2 Foster discussions of advanced strategies for using the toolbox to increase researchers' effectiveness and efficiency

Strategy 3.3 Validate the toolboxes and productivity strategies in order to reduce the do-it-yourself nature of the technology integration process

Strategy 1.1 Appreciate the systematic application of information technology in the research process

First, let's clarify our interest in technology and our motivation to spend time exploring new tools. As a hobbyist, I enjoy technology and am not opposed to spending endless hours making technology work. As an innovator, I seek to be the first to experiment with what's new. However, the primary interest for most researchers is research, not technology. Therefore, it is important to focus on the single goal: enhanced research productivity.

To appreciate the systematic application of information technology, the researcher must first have a solid understanding of the research process. This book is not an introduction to research. Readers must have sufficient experience to appreciate the demands of research tasks and be able to evaluate the usefulness of various tools. Second, readers must have insight into their own personal research agenda. Third, access to a computer is essential for implementing the ideas in this book. Fourth, it is important that researchers become proficient in using multiple programs and information resources to accomplish their work.

A fundamental concept related to technology-enhanced productivity involves the use of electronic text. Electronic text has several characteristics that make it very different from printed text. Although printed text may be reproduced by copying or distributed by faxing, other possibilities are limited due to the fact that the text appears in a certain format and size on paper. In contrast, by using my word processor to create this page, I have created electronic text. Electronic text can be modified in seconds, printed if necessary, copied and pasted into an email message and sent to a colleague, imported into a desktop publishing program to be made into a stylish handout for my class, copied and pasted into a presentation program to illustrate my lecture or workshop, and published on my home page on the web. The possibilities are endless.

The electronic scholar seeks to extend the benefits of using electronic text: to create and manipulate information across a variety of programs on the electronic desktop. The characteristics of electronic text make it the basic building block of technology-enhanced productivity. Indeed, anyone who has used a word processor recognizes the value of this tool for its power in the revision process.

To sum up strategy 1.1–to appreciate the application of information technology–the electronic scholar must have a working knowledge of a core set of desktop tools that enable her to create, manipulate, and distribute information in support of common productivity demands with greater ease and efficiency than otherwise would be possible.

Strategy 2.1 Increase our understanding of the productivity demands made on researchers

In order to discuss technology-enhanced productivity, we need to know what researchers do. For the purposes of this book, I will discuss two aspects of productivity for researchers: general productivity and research productivity.

As outlined in Table 1-3, *general productivity* involves a variety of common tasks completed by professionals, such as writing letters and locating information, that rely on the creation, storage, retrieval, and manipulation of electronic text. The core set of productivity tools used to complete these tasks is fairly generic and shared by a variety of professionals. More importantly, the toolbox used for supporting general productivity also serves as the foundation on which to build a toolbox for enhancing research productivity.

Research productivity is a term used to describe the tasks commonly completed by researchers. Table 1-4 identifies 28 tasks commonly completed by researchers and outlines the framework that will be presented in Chapters 2 through 7 based on the four general phases of the research process: designing the study, conducting the study, analyzing the data, and reporting the results. In addition, the categories of Professional Development and Research Administrator have been used to organize continuing education and managerial tasks.

Table 1-3
Common Tasks Involved in General Productivity

Creating Information
- in a text format
- in a graphic format
- in an instructional format

Communicating Information
- in print
- orally/visually
- electronically

Managing Information
- in print
- electronically

Table 1-4
28 Tasks Commonly Associated with Research Productivity

A reseacher commonly...

Professional Development
1. Seeks to enhance knowledge and skills
2. Strives to maintain current awareness

Designing the Study
3. Engages in preliminary exploration of a topic
4. Conducts an extensive review of the literature
5. Selects an appropriate methodology
6. Identifies potential sources of funding
7. Engages in collaboration
8. Prepares proposals
9. Participates in human subjects review

Conducting the Study
10. Creates materials
11. Collects data
12. Communicates with field sites

Analyzing the Data
13. Analyzes quantitative data
14. Analyzes qualitative data

Reporting the Results
15. Consults writing resources
16. Explores the visual representation of data
17. Disseminates research findings
18. Presents research results

A research administrator commonly...

Research Administrator
19. Manages information
20. Manages projects
21. Maintains financial records
22. Selects mailing systems
23. Purchases supplies
24. Purchases equipment
25. Obtains technical support
26. Facilitates travel planning
27. Prepares data recovery plans
28. Participates in technology transfer processes

Strategy 2.2 Identify appropriate tools that contribute to the completion of tasks commonly associated with research productivity

Leedy (1997) observed that the researcher's toolkit should contain the five general tools of research: (1) the library and its resources, (2) the computer and its software, (3) techniques of measurement, (4) statistics, and (5) familiarity with language (p. x). This book provides an in-depth review of item number 2 on Leedy's list: the computer and its related tools. It is our assumption here that the researcher's workstation contains a computer, appropriate software, a printer, and access to a network. A toolkit may include technology in any of the forms listed in Table 1-5. The marketplace offers researchers many choices relative to the size, shape, function, and price of tools. I encourage researchers to use a broad definition of technology when constructing their toolbox.

The use of productivity frameworks outlined in Strategy 2.1 facilitate the efforts of the electronic scholar to assemble a desktop collection of electronic tools that foster enhanced productivity. Using the list of 28 tasks commonly associated with research productivity (Table 1-4), I considered how technology might be used in the routine completion of each task. In the end, I was able to identify 78 tactics. Each tactic is supported by a number of tools to illustrate the technology tools that might be incorporated into the researcher's toolbox and used in the routine completion of a task. The taxonomy outlined in Table 1-6 offers a comprehensive view of technology-enhanced research productivity and serves to organize information on more than 750 technology tools presented in this book. Each productivity task, tactic, and tool are examined in detail in Chapters 2 through 7.

Table 1-5
Common Technologies for Enhancing Instructional Productivity

Portable Electronic Tools
    calculators
    spelling checkers
    personal organizers
    instructional learning devices

Microcomputers
    stand-alone computers
    workstations connected to a local area network
    integrated learning systems

Presentation Systems
    overhead projectors
    computer projection systems
    video cassette recorders
    videodisc players
    tape recorders
    televisions

Distance Learning Systems
    instructional programming delivered via cable
    instructional programming delivered via satellite
    two-way interactive video teleconferencing

Video Production Tools
    camcorders
    digital cameras
    video editing facilities

Communication Systems
    telephones
    voice mail systems, homework hotlines
    modems
    facsimile machines

Information Retrieval Systems
    electronic card catalogs
    CDROM based reference products

Table 1-6
Edyburn's Taxonomy of Technology-Enhanced Research Productivity

Professional Development (2 tasks, 11 tactics)
    Task #1 Enhancing Knowledge and Skills
          Tactic 1-1 Identifying resources for learning about research
          Tactic 1-2 Identifying resources for teaching about research
          Tactic 1-3 Exploring research methods and procedures
          Tactic 1-4 Participating in the research community
          Tactic 1-5 Considering career opportunities
          Tactic 1-6 Learning about anything you can imagine
          Tactic 1-7 Developing a personal library
    Task #2 Maintaining Current Awareness
          Tactic 2-1 Subscribing to electronic information services
          Tactic 2-2 Utilizing news services
          Tactic 2-3 Utilizing tools that monitor
          Tactic 2-4 Shifting from online to offline browsing

Designing the Study (7 tasks, 22 tactics)
    Task #3 Engaging in Preliminary Exploration of a Topic
          Tactic 3-1 Sending email
          Tactic 3-2 Browsing the web
    Task #4 Conducting an Extensive Review of the Literature
          Tactic 4-1 Identifying appropriate databases to search
          Tactic 4-2 Browsing web-based publications
          Tactic 4-3 Considering options for document delivery
          Tactic 4-4 Writing the literature review
    Task #5 Selecting an Appropriate Methodology
          Tactic 5-1 Selecting a research design
          Tactic 5-2 Locating resources for qualitative research
          Tactic 5-3 Locating resources for quantitative research
          Tactic 5-4 Locating resources for survey research
          Tactic 5-5 Selecting and describing subjects
          Tactic 5-6 Locating assessment instruments
    Task #6 Identifying Potential Sources of Funding
          Tactic 6-1 Subscribing to a funding alert service
          Tactic 6-2 Searching for funding sources
    Task #7 Engaging in Collaboration
          Tactic 7-1 Locating an individual
          Tactic 7-2 Using the web as a telephone
          Tactic 7-3 Using video conferencing
          Tactic 7-4 Using audiographics
          Tactic 7-5 Using work group software
    Task #8 Preparing Proposals
          Tactic 8-1 Cutting and pasting information
          Tactic 8-2 Using proposal development tools

Table 1-6 *(continued)*

Task #9 Participating in Human Subjects Review
        Tactic 9-1 Locating procedures, policies, forms

Conducting the Study (3 tasks, 11 tactics)
    Task #10 Creating Materials
        Tactic 10-1 Creating print materials
        Tactic 10-2 Using rapid prototype tools
        Tactic 10-3 Creating web-based materials
    Task #11 Collecting Data
        Tactic 11-1 Using hand-held data input systems
        Tactic 11-2 Using portable keyboards
        Tactic 11-3 Using computer-based data collection tools
        Tactic 11-4 Using unobtrusive computer-based data collection tools
        Tactic 11-5 Using web-based data collection tools
        Tactic 11-6 Using voice input
        Tactic 11-7 Using digital cameras
    Task #12 Communicating with Field Sites
        Tactic 12-1 Using email

Analyzing the Data (2 tasks, 4 tactics)
    Task #13 Analyzing Quantitative Data
        Tactic 13-1 Using generic tools for data analysis
        Tactic 13-2 Using specialized data analysis tools
    Task #14 Analyzing Qualitatitive Data
        Tactic 14-1 Using generic tools for text analysis
        Tactic 14-2 Using specialized text analysis tools

Reporting the Results (4 tasks, 9 tactics)
    Task #15 Consulting Writing Resources
        Tactic 15-1 Using ready reference tools for writers
        Tactic 15-2 Using guides to grammar, word use, and style
    Task #16 Exploring the Visual Representation of Data
        Tactic 16-1 Using generic tools for visual representation of data
        Tactic 16-2 Using creative tools for visual representation of data
        Tactic 16-3 Using specialized tools for visual representation of data
    Task #17 Disseminating Research Findings
        Tactic 17-1 Disseminating information in print formats
        Tactic 17-2 Disseminating information via the web
        Tactic 17-3 Disseminating information in electronic formats
    Task #18 Presenting Research Results
        Tactic 18-1 Using presentation software

Table 1-6 *(continued)*

Research Administrator (10 tasks, 21 tactics)
    Task #19 Managing Information
        Tactic 19-1 Creating electronic text
        Tactic 19-2 Creating databases
    Task #20 Managing Projects
        Tactic 20-1 Using project management software
        Tactic 20-2 Using time management software
        Tactic 20-3 Meeting technical requirements
    Task #21 Maintaining Financial Records
        Tactic 21-1 Using financial recordkeeping software
    Task #22 Selecting Mailing Systems
        Tactic 22-1 Mailing and tracking systems
    Task #23 Purchasing Supplies
        Tactic 23-1 Purchasing office supplies
    Task #24 Purchasing Equipment
        Tactic 24-1 Purchasing new equipment
        Tactic 24-2 Purchasing upgrades
        Tactic 24-3 Purchasing used equipment
    Task #25 Obtaining Technical Support
        Tactic 25-1 Obtaining phone support
        Tactic 25-2 Using troubleshooting tools
        Tactic 25-3 Accessing web sites that offer technical support
    Task #26 Facilitating Travel Planning
        Tactic 26-1 Making reservations
        Tactic 26-2 Exploring a city
        Tactic 26-3 Checking on the weather
        Tactic 26-4 Working while you travel
    Task #27 Preparing Disaster Recovery Plans
        Tactic 27-1 Using backup systems
        Tactic 27-2 Planning for disaster recovery
    Task #28 Participating in Technology Transfer Processes
        Tactic 28-1 Facilitating technology transfer

Strategy 2.3 Create a toolbox of electronic tools

Having closely examined general and research productivity and the tasks
associated with each, and enlarged our definition of technology, it is time to
focus on creating a toolbox of electronic tools. Confronted with the opportunity
to purchase a new computer and fill the hard drive with all sorts of new products, what should we select? (Indeed, it is no longer difficult to fill a hard drive,
no matter how big it is!). The challenge is to align appropriate tools with common tasks.

Figure 1-1 is a worksheet for the electronic scholar to use when building an
electronic toolkit to support general productivity. Four clusters of productivity
demands are outlined: Creating Information, Communicating Information,
Managing Information, and Using Specialized Tools. Readers should use this
worksheet to record on a regular basis the names of products currently used.
Rows that are left blank illustrate areas you may need to potentially explore. Use
your worksheet to discuss your needs with others. You may discover that you
do not have a need for certain types of tools based on the kind of work that you
do. You may also find that some of your existing tools have features that you've
yet to use regularly. Of course, you may also discover the need to go shopping!

Once you have developed the general productivity toolkit, you can build the
research productivity toolkit. Figure 1-2 provides a worksheet for the electronic
scholar to use when building an electronic toolkit to support research productivity. The 28 common tasks are clustered in the four phases of the research process
along with sections on professional development and research administration.
You should use this worksheet to record your discoveries as you complete
Chapters 2 through 7.

Figure 1-1
General Productivity Worksheet

Task: Creating Information	Possible Tools
in a text format	
word processor	
electronic writing aids	
in a graphic format	
paint and draw	
graph and chart	
desktop video	
in an instructional format	
text and graphics	
exam/study guides	

Task: Communicating Information	Possible Tools
in print	
desktop publishing	
orally/visually	
presentation software	
electronically	
email	
web browser	
work group software	

Task: Managing Information	Possible Tools
in print	
labels, forms	
electronically	
databases	
spreadsheet	

Task: Using Specialized Tools	Possible Tools
desktop	
calendars, planners	
utilities	
file conversion	
general utilities	
virus protection	

14

Figure 1-2
Research Productivity Worksheet

For each numbered task, record the web addresses of technology tools you find useful for your research.

Professional Development

1. **Enhancing knowledge and skills**

2. **Maintaining current awareness**

Designing the Study

3. **Engaging in preliminary exploration**

4. **Conducting a review of the literature**

5. **Selecting an appropriate methodology**

6. **Identifying sources of funding**

7. **Engaging in collaboration**

8. **Preparing proposals**

9. **Participating in human subjects review**

Figure 1-2 *(continued)*

Conducting the Study

10. Creating materials

11. Collecting data

12. Communicating with field sites

Analyzing the Data

13. Analyzing quantitative data

14. Analyzing qualitative data

Reporting the Results

15. Consulting writing resources

16. Exploring the visual representation of data

17. Disseminating research findings

18. Presenting research results

Figure 1-2 *(continued)*

Research Administrator

19. Managing information

20. Managing projects

21. Maintaining financial records

22. Selecting mailing systems

23. Purchasing supplies

24. Purchasing equipment

25. Obtaining technical support

26. Facilitating travel planning

27. Preparing disaster recovery plans

28. Participating in technology transfer processes

Next Steps

Several strategies for becoming an electronic scholar have yet to be considered:

Strategy 3.1 Facilitate training in the use of the toolbox

Strategy 3.2 Foster discussions of advanced strategies for using the toolbox to increase researchers' effectiveness and efficiency

Strategy 3.3 Validate the toolboxes and productivity strategies in order to reduce the do-it-yourself nature of the technology integration process

However, these concerns are more properly addressed after you have had hands-on experience and deeper exploration of the possibilities. As a result, we'll pause here and continue our discussion on toolkits and productivity strategies in Chapter 8.

In preparation for Chapter 8, complete the following activities:

Activity #1: The checklist in Table 1-7 defines skills involved in navigating the world wide web. If you are a beginner, use this checklist as a guide to developing your skills. If your skills are in the intermediate or advanced range, you are ready to take advantage of the information in this book; use your talents to assist your colleagues in getting started.

Activity #2: Use the worksheet in Figure 1-1 to document your existing general productivity toolkit. Discuss your toolkit with others to identify additional products that may be useful in your work.

Activity #3: Use the worksheet in Figure 1-2 to record your discoveries as you complete Chapters 2 through 7.

Table 1-7
World Wide Web Self-Assessment Skill Checklist

Name _____

Type of Computer (circle one):   Windows   Macintosh

Type of Browser (circle one):
    America Online   Microsoft Internet Explorer   Netscape Navigator

Check your skill level prior to this course in the Pre column. At the conclusion of the course, check your skill level in the Post column.

Pre	Post	AWARENESS
___	___	I have heard of the world wide web.
___	___	I know someone who has "surfed the web."
___	___	I have used a web browser to "surf the web."
___	___	I know where/how I could access the web at home or work.

Pre	Post	BASIC SKILLS
___	___	Demonstrate the ability to point and click.
___	___	I can locate Netscape on the computer hard drive and initiate the program.
___	___	I know how to recognize a home page.
___	___	I know how to enter a www address into Netscape.
___	___	Demonstrate the ability to use scroll bars.
___	___	I can recognize the visual cues indicating a "link."
___	___	Demonstrate the ability to select and access a link.
___	___	Demonstrate the ability to use the "Back" button.

Pre	Post	INTERMEDIATE SKILLS
___	___	Demonstrate the ability to conduct a search on the web.
___	___	I know how to obtain "copies" of selected information using the options for (a) Save As, (b) Mail Document, and (c) Print.
___	___	I have started collecting addresses of useful web sites.

Pre	Post	ADVANCED SKILLS
___	___	I feel comfortable in teaching others how to navigate the web.
___	___	Demonstrate the ability to create a home page.
___	___	Demonstrate the ability to create working links.
___	___	Demonstrate the ability to copy selected portions of html code from a page and insert this code into a personal home page (e.g., graphic, animation, format, etc.).

References

Brent, E.E., & Anderson, R.E. (1990). *Computer applications in the social sciences.* New York: McGraw-Hill.

Hammer, M., & Champy, J. (1993). *Reengineering the corporation.* New York: Harper Business.

Harrison, T.M., & Stephen, T. (1996). *Computer networking and scholarly communication in the twenty-first century university.* New York: State University of New York Press.

Kelle, U. (1995). *Computer aided qualitative data analysis: Theory, methods, and practice.* Thousand Oaks, CA: Sage.

Kennedy, P. (1993). *Preparing for the twenty-first century.* New York: Random House.

Lawrence, J.S. (1984). *The electronic scholar: A guide to academic microcomputing.* Norwood, NJ: Ablex.

Leedy, P.D. (1997). *Practical research: Planning and design.* Upper Saddle River, NJ: Merrill/Prentice-Hall.

Naisbitt, J. (1982). *Megatrends.* New York: Warner Books.

Peppers, D., & Rogers, M. (1997). *Enterprise one to one: Tools for competing in the interactive age.* New York: Doubleday.

Quality Education Data. (1997). *1997-98 Technology purchasing forecast.* Denver, CO: Author.

Stewart, T.A. (1997). *Intellectual capital: The new wealth of organizations.* New York: Doubleday/Currency.

Tesch, R. (1990). *Qualitative research: Analysis type and software tools.* Bristol, PA: Falmer Press.

CHAPTER 2
Professional Development

Earlier in the century, a graduate degree provided extensive training in research methods–training that could be expected to last a career. Now, as the century draws to a close, the longevity of knowledge is often discussed in half-lives, or the amount of time it takes for half of the information in a field to be rendered obsolete. Some estimates indicate that the half-life of information is 3 to 5 years. Researchers must therefore make a significant commitment to ongoing professional development, which may focus on, for example, learning about a new analysis procedure or establishing a new direction in one's research program. The purpose of this chapter is to examine two common tasks involved in the professional development of researchers: enhancing knowledge and skills, and maintaining current awareness.

Task #1 Enhancing Knowledge and Skills
Tactic 1-1 Identifying resources for learning about research
Tactic 1-2 Identifying resources for teaching about research
Tactic 1-3 Exploring research methods and procedures
Tactic 1-4 Participating in the research community
Tactic 1-5 Considering career opportunities
Tactic 1-6 Learning about anything you can imagine
Tactic 1-7 Developing a personal library

Task #2 Maintaining Current Awareness
Tactic 2-1 Subscribing to electronic information services
Tactic 2-2 Utilizing news services
Tactic 2-3 Utilizing tools that monitor
Tactic 2-4 Shifting from online to offline browsing

Task #1: Enhancing Knowledge and Skills

Professionals in any field are confronted with the challenges of mastering the knowledge and skills necessary initially to enter their profession and subsequently to maintain and expand their knowledge and skills through continuing education. This section describes seven tactics to assist researchers in utilizing technology to enhance their knowledge and skills. Each tactic includes a list of pertinent web site addresses.

Tactic 1-1 Identifying resources for learning about research

The web features a variety of resources for learning about research. These sites are useful for students preparing for their initial entry into the profession and for researchers interested in learning about new methodologies or analysis techniques.

Statistics Every Writer Should Know
http://nilesonline.com/stats/
> Designed by Robert Niles, this site was created to describe key statistical concepts in plain English to help those who need to write about statistics.

Statistics Glossary
http://www.stats.gla.ac.uk/steps/glossary/index.html
> A glossary of terms commonly used in statistics thematically organized, developed by Stuart G. Young of the University of Glasglow. An alphabetical index is also available.

Office of Educational Research and Improvement
http://www.ed.gov/offices/OERI
> A description of the government office with primary responsibility for educational research. Includes an extensive collection of links to federal programs and resources. See Figure 2-1.

SRM Database of Social Science Research Methodology
http://www.scolari.com/SMR.HTM
> A Windows-compatible CDROM with more than 40,000 literature references in social science methodology from Sage Publications. Search by author, publisher, language, publication year, or use a term from the 1,166-item index. Full-text searching is also possible.

Figure 2-1
Web Page of the Office of Educational Research and Improvement,
U.S. Department of Education

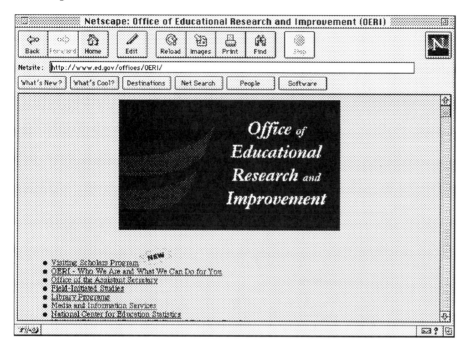

Tactic 1-2 Identifying resources for teaching about research

Faculty involved in teaching courses in educational research can find a wealth of information online. Many resources may be located through links on home pages of statistics departments, personal home pages of individual instructors, and professional associations. Increasingly, not only are data sets available online, but also analysis and graphing tools that afford instructors the opportunity to use the web in the classroom and teach interactively.

Web sites maintained by individual instructors

Statistical Instruction Internet Palette (SIIP)
http://seamonkey.ed.asu.edu/~behrens/

SIIP was created by Professor Behrens and colleagues from the College of Education at Arizona State University to enhance a series of research methods resources. Visitors will discover an integrated set of tools and resources thematically organized: equation gallery, collaboration gallery, classroom gallery, data gallery, www data, computing studio, statopedia, and graphing studio.

Statistics on the Web
http://www.execpc.com/~helberg/statistics.html
A huge site devoted to teaching statistics. Developed and maintained by Professor Helberg, University of Wisconsin-Milwaukee.

The Knowledge Base: An Online Research Methods Textbook
http://trochim.human.cornell.edu/kb/kbhome.htm
A complete electronic research textbook. Developed and maintained by Professor William M.K. Trochim of Cornell University, this site is organized into three topics: Introduction to Research; Sampling, Measurement, and Design; and Data Analysis. A very impressive work.

World Lecture Hall
http://www.utexas.edu/world/lecture/
Looking to find colleagues that teach courses similar to yours? This site has an index of courses by discipline with links to the instructor's home page to allow you to view syllabi, assignment guides, web resources, and so on. Developed and maintained by the University of Texas-Austin.

Web sites maintained by statistics departments

University of Florida, Department of Statistics Home Page
http://www.stat.ufl.edu
An extensive collection of resources developed and maintained by the Department of Statistics at the University of Florida. Home of the Statistics Virtual Library. Provides an international list of departments, divisions, and schools of statistics.

UCLA Statistics Home Page
http://www.stat.ucla.edu
A well-designed home page developed and maintained by the UCLA Statistics Department features information on courses, faculty, and so on. Be sure to check out the services and archives as well as the UCLA Electronic Textbook and Statistics Toolbox.

StatLib
http://www.stat.cmu.edu/
An extensive site maintained by the Statistics Department of Carnegie Mellon University. StatLab provides statistical software, datasets, and information by email, FTP, and www.

Web sites maintained by professional associations

American Statistical Association, Section on Statistical Education
http://www.stat.ncsu.edu/stated/

The section on Statistical Education is a special interest group within the American Statistical Association. Includes links to the journal, *Journal of Statistics Education*, newsletter, statistics teacher network, undergraduate data analysis contest, data sets, and graduate programs.

STEPS: Statistical Education through Problem Solving
http://www.stats.gla.ac.uk/steps/index.html

A consortium has developed a series of problem-based modules to support the teaching of statistics in biology, business, geography, and psychology. Features free software available for downloading.

STATS: Statistical Thinking with Active Teaching Strategies
http://stats.dickinson.edu

A project supported by the National Science Foundation and the Mathematical Association of America. Features online resources and information on regional faculty development workshops. Designed to use active teaching strategies for the improvement of students' thinking and statistical problem solving.

Instructional products

ActivStats
http://www.datadesk.com

Located on the Data Desk home page, this program is a complete multimedia presentation of material commonly found in an introductory statistics course. The interactive learning system integrates video, simulation, animation, narration, video, interactive experiments, and web access into a rich learning environment.

K-12 Statistics
http://www.mste.uiuc.edu/stat/stat.html

As the title suggests, this site offers resources for teaching statistics in grades K-12. Organized around the National Council of Teachers of Mathematics (NCTM) math standards, it provides practical activities for teaching statistical concepts.

Data sets

Statistical Resources on the Web
http://www.lib.umich.edu/libhome/Documents.center/stats.html
A huge collection of statistics databases available on the web. This site is maintained by the University of Michigan Documents Center.

Fedstats
http://www.fedstats.gov/
Databases full of statistics and information from federal agencies.

U.S. Census Bureau
http://www.census.gov
Data held in the public domain collected and disseminated by the United States Census Bureau. Provides data extraction tools and tons of data on almost any topic.

The State of the Nation's Cities: A Comprehesive Database on American Cities and Suburbs
http://policy.rutgers.edu/cupr/SoNC.htm
This well-known database features over 3,000 variables concerning employment and economic development, demographic measures, housing and land use, income and poverty, fiscal conditions, and a variety of other health, social, and environmental indicators for 77 American cities and suburbs. Maintained by Rutgers University's Center for Urban Policy Research under contract with the U.S. Department of Housing and Urban Development.

Tactic 1-3 Exploring research methods and procedures

Despite the fact that many research procedures are common among the sciences, observing how other disciplines solve problems is often inspirational. This tactic provides the opportunity to explore new ways of thinking about research. Your investment in exploring the sites listed in this section has the potential to offer significant dividends for your work, particularly in the areas of data collection and dissemination of information.

Life sciences

National Human Genome Research Institute
http://www.nhgri.nih.gov/
Home page with links to the large science project devoted to gene mapping. An awesome example of collaboration, publication of data sets, and the development of powerful analysis tools.

National Center for Biotechnology Information
http://www.ncbi.nlm.nih.gov/
Another example of a large science project (building databases of DNA, RNA, amino acid, and 3-D structures of protein molecules) and the coordinated effort to share information about research in progress.

Visible Human Project
http://www.nlm.nih.gov/research/visible/visible_human.html
The National Library of Medicine is creating complete, anatomically detailed, 3-D representations of the male and female human body with images collected at one-millimeter intervals. The long-term goal of the project is to produce a system of knowledge structures that will transparently link visual knowledge forms to symbolic knowledge formats.

Plants National Database
http://plants.usda.gov/plants/
This site claims to be the "single source of general information about the plant kingdom." Visitors will find little to dispute this claim. Features a database, a photo gallery of 1,000 plant images, as well as links to other databases and resources. Maintained by the U.S. Department of Agriculture.

Project Feeder Watch
http://www.ornith.cornell.edu
Sponsored by the Cornell Lab of Ornithology, an online reporting system for 12,000 bird watchers to collect long-term data on winter bird populations. Features a section on Citizen Science and the Bird of the Week.

Geology

National Ice Core Laboratory
http://gust.sr.unh.edu/NICL/
Photos and data from the National Ice Core Laboratory illustrate the increasingly collaborative nature of research.

Education

What do we need to know to improve learning?
U.S. Department of Education
http://www.ed.gov/offices/OERI/RschPriority/learning.html

Interesting example of an online survey. Visitors are encouraged to provide feedback by responding to specific questions. Data is gathered into an electronic database for use in subsequent policy-making decisions.

Educational Resource Network on the Web
http://www2.ernweb.com/ernweb

Developed by the publishers of the print newsletter, *Educational Research Newsletter,* this site is a community of people concerned with education. Lots of resources and links. Selected summaries from their newsletters. Request a free sample copy.

Intenet Resources for Institutional Research
http://highered.org/links/linkmap.html

A comprehensive site devoted to research in higher education. A "must visit:" site for instututional researchers, faculty, students, and others interested in higher education.

History

Natural History Web
http://nmnhgoph.si.edu/nmnhweb.html

Explore the Smithsonian National Museum of Natural History. A well-designed site with access to research and collections in seven scientific disciplines.

World War II Poster Collection
http://www.library.nwu.edu/govpub/collections/wwii-posters/

An image database in which you can search or browse more than 200 posters related to World War II. Developed and maintained by the Government Publications Department of the Northwestern University Library, this site offers an inspirational perspective on the digitalization of images and the value of these tools for researchers.

The Papers of George Washington
http://www.virginia.edu/gwpapers/

Read and view samples of George Washington's correspondence. An ambitious project that seeks to place all of Washington's papers online.

The Getting Word Oral History Project
http://www.monticello.org/gettingword

A project by the Thomas Jefferson Memorial Foundation at Monticello, this site describes the procedures used in this oral history project and illustrates the results of the project. Follow the fate of Thomas Jefferson's slaves and their descendants.

Just the Arti-facts
http://www.chicagohs.org/AOTM/jta.html

This site was created by the Chicago Historical Society as a means of enabling teachers and students to explore artifacts from its extensive collection. Images of selected Chicago history artifacts, stories documenting their historical significance, photos, and audio selections are provided for each monthly feature.

Interdisciplinary

National Science Foundation
http://www.nsf.gov/

The National Science Foundation sponsors programs in a variety of disciplines, including education. Provides access to science statistics, databases, and program descriptions.

National Academy of Science
http://www.nas.edu

A common home page for four organizations: National Academy of Science, National Academy of Engineering, Institute of Medicine, and the National Research Council. The front page serves as a daily news update with topical links to information at the site.

SCILINK, Inc.
http://www.scicentral.com

Directory of 50,000 sites of more than 120 specialties in science and engineering. New efforts to link K–12 and educational researchers.

Figure 2-2
Linguisitic Resources on the Internet
Used with permission of Summer Institute of Linguistics.

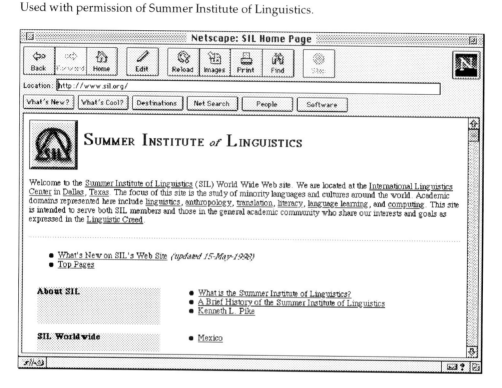

Social sciences

Linguisitic Resources on the Internet
http://www.sil.org/

This site monitors the study of minority languages and cultures around the world with specific emphasis on linguisitics, anthropology, translation, literacy, language learning, and computing. Developed and maintained by the Summer Institutes of Linguistics at the International Linguistics Center in Dallas, Texas. See Figure 2-2.

Transactional Records Access Clearinghouse (TRAC)
http://www.trac.syr.edu

A comprehensive source of information about the activities of federal law enforcement and regulatory agencies. Provides highly detailed information on four subjects: IRS; Bureau of Alcohol, Tobacco, and Firearms; Drug Enforcement Administration; and FBI. TRAC is a data gathering, data research, and data distribution organization associated with Syracuse University.

Research Techniques

Bee Alert
http://grizzly.umt.edu/biology/bees

This site lets visitors literally visit with Professor Bromenshenk of the University of Montana as he conducts his research on bees. Features a "bee-cam," extensive facts about bees, real-time and archived data from the beehives he monitors. An interesting experiment illustrating what happens when we open our research laboratories to anyone who cares to stop by.

Crime and Justice Electronic Data Abstracts
http://www.ojp.usdoj.gov/bjs/dtdata.htm

A massive database updated by the U.S. Department of Justice Bureau of Justice Statistics, organized into 55 subjects covering crimes and arrests, criminal justice, and general demographics. The data are aggregated from a variety of published sources and are available in Lotus spreadsheet format to facilitate data analysis.

The Genetic Algorithms Archive
http://www.aic.nrl.navy.mil/galist/

This site is a repository for information related to research in genetic algorithms. Maintained by Alan C. Schultz of the Navy Center for Applied Research in Artifical Intelligence. The archive is a valuable concept for researchers in other disciplines.

Digital Library
http://www.archive.org

This site describes an in-progress project to archive the whole web. Really! The goal is to build a digital library that will support future research about the web.

Apple Classrooms of Tomorrow
http://www.research.apple.com/go/acot/

Home page describing the research from Apple Classrooms of Tomorrow (ACOT). Provides a link to ACOT research reports available in Adobe Acrobat pdf format.

Biodiversity and Conservation
http://darwin.bio.uci.edu/~sustain/bio65/Titlpage.htm

This site features a hypertext book developed by Peter J. Bryant, a Professor of Developmental and Cell Biology at the University of California, Irvine, for his class in Biological Conservation. A comprehensive introduction to the topic featuring six chapters. Makes extensive use of hyperlinks. Each chapter also contains illustrations, photos, charts, and a webliography.

AlphaWorks
http://www.alphaworks.ibm.com

A web-based community for cutting-edge Internet research and technology. Registered users can access real-time demos and online prototypes, comment on, download, and experiment with early-stage technologies. This site has helped commercialize a number of products.

Disseminating research information beyond the research community

3-D Insects
http://www.ento.vt.edu/~sharov/3d/3dinsect.html

Created by Professor Alexei A. Sharov, a research scientist in the Department of Entomology at the University of Virginia Polytechnic Institute and State University, this site features QuickTime movies of 14 insects with links to other web resources to learn more about these creatures. An engaging visit!

Food Finder
http://www.olen.com/food/

Using information from the *Fast Food Handbook* published by the Minnesota Attorney General's Office, this site is an interactive tool. Food Finder allows users to search any of 19 fast food restaurants by food names, maximum calories, percent calories from fat, maximum sodium, fat, and cholesterol. Submit your search by clicking on the button, "fire up the deep fryer." A fun and informative site for considering the effect of your next fast food meal on your health.

Consultant: A Diagnostic Support System
for Veterinary Medicine
http://www.vet.cornell.edu/consultant/consult.asp

This site, designed as a tool to aid trained veterinarians, features a database of approximately 500 signs and symptoms, about 4,000 diagnoses, and more than 10,000 literature references to diseases in 8 small- and large-animal groups. Users can search by sign keyword or diagnosis. Diagnosis includes a description, the species it applies to, a list of signs, reference to pertinent literature, and links to related web sites when available. Created and maintained by Professor Maurice E. White of the College of Veterinary Medicine at Cornell University.

Why Files
http://whyfiles.news.wisc.edu

> Designed to engage the layperson, features an in-depth exploration of the science behind the headlines. Focuses on biology, environmental science, health, physcial science, social science, sports, and technology.

American Memory
http://lcweb2.loc.gov/ammem/ammemhome.html

> A project of the Library of Congress to create Historical Collections for the National Digital Library. American Memory contains a wealth of resources, including: learning guides, photos, sound files, and hyperlinks. An inspirational vision of the future of virtual museum collections.

Tactic 1-4 Participating in the research community

Experienced researchers participate in a network with other researchers where ideas and resources are exchanged. Because new researchers often find it difficult to become connected within this "invisible college," the purpose of this tactic is to consider ways to help new researchers participate in the research community. In addition to resources for matching research interests, tools are provided to assist in joining a professional association and identifying appropriate professional conferences to attend.

Community of Science
http://www.cos.com

> Join the Community of Science by completing a profile of your research interests. Your profile will be added to the Expertise database. Search for other researchers with similar research interests. If you still need more incentive to motivate you to join, be advised that once your profile is in the system you will receive email alerting you to funding opportunities that appear to match your research interests.

U.S. Department of Education Research Priorities
http://www.ed.gov/offices/OERI/RschPriority/

> This policy document outlines the current and future research priorities for the U.S. Department of Education. Useful to review in light of your own research agenda.

Figure 2-3
Home Page of the American Educational Research Association
Used with permission of American Educational Research Association.

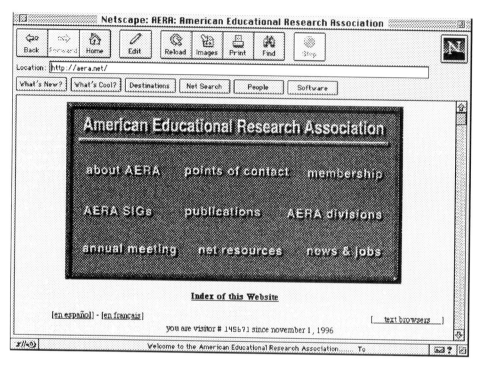

Joining a professional association

University of Waterloo Electronic Library
Scholarly Societies Project
http://www.lib.uwaterloo.ca/society/overview.html

> A collection of links to information about academic societies around the world. Search more than 1,100 scholarly societies by subject guides, search engines or society name. Also features a meeting conference announcement list.

Education Associations and Organizations
http://www.ed.gov/EdRes/EdAssoc.html

> An extensive list of professional organizations and links to their web sites. Maintained by the U.S. Department of Education.

American Educational Research Association
http://aera.net

> The premier professional organization for researchers studying education. See Figure 2-3.

Attending professional conferences

ERIC Calendar of Education-Related Conferences
http://www.aspensys.com/eric/cgi/ccal.cgi
Search for education-related conferences of potential interest by subject, sponsor, city, state, or dates.

Tactic 1-5 Considering career opportunities

From time to time, new and experienced researchers are likely to consider opportunities that will enhance their careers. As you might expect, the web can be an invaluable tool for locating position announcements, learning more about a specific university, and obtaining information on a host of relocation details.

Position announcements

Chronicle of Higher Education
http://chronicle.com
Search hundreds of university position announcements published each week in *The Chronicle of Higher Education.*

Academic Employment Network
http://www.academploy.com
Search for teaching and other school-related positions or post an announcement of a position you have available. Includes extensive resources and links.

Educational Placement Service
http://www.educatorjobs.com
The largest teacher placement service in the United States. Accredited by the National Association of Teacher Agencies.

The Monster Board
http://www.monster.com
For job seekers and employers. Check out this site and see why its slogan claims that is is "the premier career hub on the www." Be sure to check out "Swoop," your personal search agent.

Fellowships

Office of Scientific and Engineering Personnel, Fellowship Programs Unit
http://www2.nas.edu/fo/
Sponsored by the National Research Council, this site describes several government agencies that sponsor a variety of predoctoral, dissertation, and postdoctoral fellowship programs on an annual basis. Also has links to other fellowship resources on the web.

The Fulbright Scholar Program
http://www.usia.gov/education/srusfulb.htm
This program offers U.S. faculty, professionals, teachers, and students the opportunity to conduct research, teach, or study abroad and to make a major contribution to the growth of mutual understanding among countries and individuals. The program also brings foreign nationals to the United States to study, teach, and pursue research.

Specific colleges and universities

The web offers great resources about specific colleges and universities to prepare for that interview! The AT&T site listed below provides an index and links to the actual college or university home page. The other sites are primarily designed to assist students in selecting a college but they provide useful descriptions of the college or university, programs, and community, and in some cases, offer virtual tours.

AT&T Universal College Home Page Index
   http://www.att.com/ucs/college/col_indx.html
CollegeView
   http://www.collegeview.com
College Edge
   http://CollegeEdge.com
College Guides and Aid Home Page
   http://sourcepath.com
CollegeNet
   http://www.collegenet.com
Peterson's Education Center
   http://www.petersons.com/
Virtual Campus Tours
   http://www.campustours.com/

Relocating

After you have decided to accept a new position, the following sites will assist you in locating housing. Each site offers search tools to assist in locating the type of housing you desire in a given area. Some sites offer street maps and photos of the houses in addition to the general description and contact information.

Realtor.Com Home Search
   http://www.realtor.com
Cyberhomes
   http://www.cyberhomes.com
HomeScout
   http://www.homescout.com
Abele Owners' Network
   http://www.owners.com

Relocation Calculators
http://www.homefair.com/home/
   Before you accept a new position, check into this site. Includes a wealth of information, calculators (see the salary calculator), guides, and references to assist you in your decision about moving from one city to another.

Tactic 1-6 Learning about anything you can imagine

This tactic explores the need to locate information on almost any topic. One of the first tools users learn to use is a search engine. There are many to choose from and each has its own strengths and weaknesses. Many other resources are also available to support your learning.

Directories

Sites that feature directories are organized by subject, which makes browsing easy. To be considered for inclusion in a directory, a site submits its address; if the editors feel that the site makes a unique or valuable contribution, it is added to the list. Directories are therefore not necessarily comprehensive.

A2Z
   http://a2z.lycos.com/
Excite
   http://www.excite.com/
Goto.com
   http://www.goto.com

Magellan
    http://www.mckinley.com/
PointCom
    http://www.pointcom.com/
Yahoo!
    http://www.yahoo.com/

Search engines

Search engines capture information about other sites on the web and organize
them by keyword. The interface requires that you enter a keyword which is then
matched with those in the index. The results are displayed in a form that attempts to rank relevance. Search engines, listed below, are especially useful
when you can formulate your information needs in a single word or phrase. In
contrast, directories are more useful than a search engine for initally accessing a
topic that you don't know much about.

Yahoo!
    http://www.yahoo.com/
Excite
    http://www.excite.com/
Lycos
    http://www.lycos.com/
HotBot
    http://www.hotbot.com/
AltaVista
    http://www.altavista.digital.com
InfoSeek
    http://www.infoseek.com/
WebCrawler
    http://www.webcrawler.com

Search Engine Watch
http://www.searchenginewatch.com/
    An authoritative site that monitors search engine developments. The section
    on search engine status reports is particularly valuable for end-users and
    webmasters for evaluating the strengths of various search engines.

Figure 2-4
The All in One Search Page
Used with permission of William Cross.

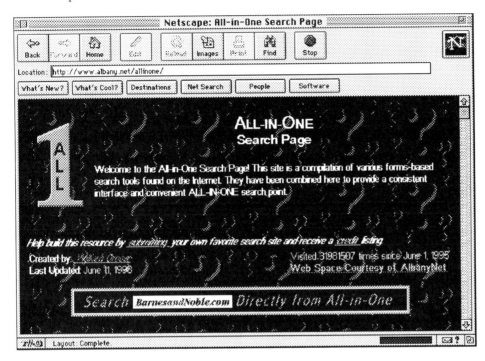

Metasearch sites

For a comprehensive search, it is often more productive to enter search terms
into multiple search engines. Metasearch sites, such as those listed below,
streamline this process. When you submit your keyword search to a metasearch
site, it conducts the search using multiple search engines at the same time.

The All in One Search Page (See Figure 2-4)
    http://www.albany.net/allinone/
AlphaSearch
    http://www.calvin.edu/Lib_Resources/as/
AccuFind
    http://www.nln.com/
Find-It
    http://www.itools.com/
Internet Sleuth
    http://www.isleuth.com

MetaCrawler
   http://metacrawler.cs.washington.edu/index.html
Metasearch
   http://metasearch.com
Search.Com
   http://www.search.com
Starting Point
   http://www.stpt.com

Daily lessons

A number of sites offer learning opportunities in daily bit-size portions. If this form of learning appeals to you, simply designate one of these sites as your home page. Then, when you launch your web browser each day it will automatically connect you to the site. A great way to begin each day!

Cool Site of the Day
http://cool.infi.net

Check this site daily and your "cool IQ" will rise dramatically as you expand your vision of what's possible on the web.

Dummies Daily
http://www.dummiesdaily.com

Yes, you guessed it! The folks who brought you the *Dummies* series of books, now offer a daily briefing. Subscribe to one or more of the Dummies Daily 13 electronic newsletters. Sent to you via email, the daily newsletter has computer tips, hints, and tricks to improve your mastery over the technology.

What's Going On
http://www.whatsgoingon.com

Descriptions of what's happening today in a comprehensive list of topics. If you have specific interests, conduct a search using the Event-O-Matic.

This Day in History
http://www.historychannel.com/historychannel/thisday

As the title suggests, review all sorts of information about events that happened on this day in history.

Frequently asked questions

Frequently asked questions (FAQs) are a convenient way to locate information about a topic. FAQs are available on most sites. Several directories and search engines index FAQs, making it easy to locate an answer to your question when you are not sure where to turn.

FAQ Findere
   http://ps.superb.net/FAQ
FAQ Search Engine
   http://www.lib.ox.ac.uk/search/search_faqs.html
FAQ Finder
   http://faqfinder.cs.uchicago.edu

Opportunities for self-directed learning

Ask an Expert
http://www.askanexpert.com/askanexpert/
   Access to more than 300 websites and experts worldwide to answer your questions in 12 categories. Interesting!

Learn2
http://Learn2.com
   This delightful site provides "2torials" that teach you practical skills in just a few minutes.

Online classes

Increasingly, the web is viewed as a way to deliver instruction. The following sites offer online instruction.

Learn the Net
http://www.learnthenet.com
   Free web-based Internet classes on a variety of Internet topics including: getting started, Internet basics, email, multimedia, newsgroups, conferencing, and web publishing. Especially valuable since the site is multilingual.

LearnItOnline
http://www.learnitonline.com
Ziff-Davis's online learning center. Register and try several lessons free.
Very affordable self-study tutorials.

ZD Net University
http://www.zdu.com
All the courses you want for $4.95 per month. Ziff-Davis Net University
classrooms are moderated message boards that work similar to newsgroups.
Each week the instructor posts lessons and assignments. Earn continuing
education credits for your work. Classes tend to have a technical focus.

Roadmap '96
http://rs.internic.net/nic-support/roadmap96
The Internet Workshop by Patrick Crispen is a free online workshop. Classes
begin every other week. Check out the syllabus. Twenty-seven lessons are
designed to be completed in six weeks.

DigitalThink
http://www.digital.com
A web-based training company that creates and distributes training courses
via the Internet in computer programming, multimedia, and entry-level
Internet topics.

Real Education
http://www.realeducation.com
Real Education is a company that specializes in creating web-based online
universities and training centers.

CASO's Internet University
http://www.caso.com/iuhome.html
A vast listing of articles, courses, and study resources. Links to more than
700 liberal arts courses in conjunction with colleges and universities around
the United States. Developed and maintained by Cape Software, Harwich
Massachusetts.

Lifelong Learning
http://www.geteducated.com/
The Adult Education and Distance Learning Resource Center offers links to
accredited distance education programs. Developed and maintained by
Lifelong Learning, Waterbury, Vermont.

The MASIE Center
http://www.masie.com
> A collection of resources on technology and learning. The MASIE Center is described as a technology and learning thinktank, located in Saratoga Springs, New York.

Other methods of learning

Video Professor
http://www.videoprofessor.com
> The Video Professor features video and CDROM-based training especially designed to assist users in learning new software products.

Tactic 1-7 Developing a personal library

A researcher's personal library is invaluable for supporting scholarship. While it is unlikely that electronic resources will overtake our desire for filling book-shelves and filing cabinets in the near future, it is important to understand the role of electronic resources. The resources supporting this tactic focus on gathering, storing, and retrieving electronic text.

Electronic text

A key component of the electronic library is electronic text. Text may be created on a word processor, scanned into the computer, or collected from another site. Once information is digitized, it can be copied, pasted, printed, or published. A key consideration for the electronic scholar is how to digitize information (retype it, scan it, capture it from a web page, save an email message) and then how to store it in a convenient location (disk, hard drive, web page, email directory).

Filing strategies

Although the task of filing is seldom high on anyone's to-do list, making time to regularly store materials for subsequent retrieval is critical. In the book, *Information Anxiety* (1989, New York: Doubleday) Richard Wurman argues that all organizational systems are based on one of the following principles: category, time, location, alphabet, or continuum. Computer directories can be used to sort your files based on their alphabetical title and the date/time the file was last accessed. Folders are used to store files categorically. Storing materials by location is increasingly common as a result of the world wide web. Of course, the best filing strategies are the ones that facilitate future retrieval.

Databases

Databases are programs designed to organize and store information. Although they are extremely powerful and flexible, most people have yet to create databases to track personally meaningful information; they simply do not have the time necessary to construct the database, and enter and maintain the information. Nonetheless, researchers should consider the practical value of a database: it can help them to manage large amounts of information and create timely reports for monitoring the information.

Microsoft Access
> **http://www.microsoft.com/access/**

FileMaker Pro
> **http://www.filemaker.com**

Paradox
> **http://www.corel.com**

Bibliographic databases

Over time a researcher builds a personal knowledge base of the literature. Bibliographic databases are specialized databases that record abstracts and citations discovered in the review of literature. By storing information in a bibliographic database, the researcher can manage reference lists and print selected items in a variety of formats.

EndNotes
> **http://www.niles.com**

ProCite, Reference Manager
> **http://www.risinc.com**

Citation 7.1
> **http://www.oberon-res.com**

Papyrus
> **http://www.teleport.com/~rsd/**

Regular backups

Losing data is one of the biggest fears of a researcher. The sensitivity of storage media requires that backups of your critical data be made on a regular schedule. (See Tactic 27-1)

Synchronizing files

An essential development in the process of collaborative workgroups is remote access for fileservers, which enables all members of the research team to access information that is stored in a single site. Initially, however, research data and reports are often generated on a single machine and must be transferred to another computer. A problem that arises is how to determine which version of a file is the most current.

One useful tool for synchronizing files is the Windows 95 Briefcase utility. To use this application simply indicate the files that you want to keep synchronized. Then, after you have made edits in a file, Briefcase will make copies using a cable or a floppy disk to keep your disk or laptop up-to-date.

The option of synchronizing files is also available with most hand-held computers (see Tactic 11-1), thus enabling you to collect data in the field and upload the information to your office computer.

Task #2: Maintaining Current Awareness

A significant challenge for professionals in all fields is staying current with new developments. Although technology has contributed to the problem of information overload, new technology tools can assist you in monitoring developments in areas of personal or professional interest. In this section, four tactics are described for utilizing technology to manage the ever-increasing amount of information.

Tactic 2-1 Subscribing to electronic information services

Electronic information services distribute new information to subscribers electronically rather than on paper. Given the time lag involved in printing and mailing, time-sensitive information is increasingly being distributed electronically. For electronic scholars, electronic text is a tremendous asset because selected materials can be saved as part of their personal library. The tools in this section describe current awareness reports and the use of listservs or mailing lists.

Social Research Update
http://www.soc.surrey.ac.uk/sru/About_sru.html

A quarterly publication from the Department of Sociology at the University of Surrey, United Kingdom. Each update is designed to assist social science researchers stay up-to-date.

Figure 2-5
The Scout Report Home Page
Copyright Internet Scout Project, University of Wisconsin-Madison, 1994-1998.
http://scout.cs.wisc. edu/

The Scout Report
http://scout.cs.wisc.edu/scout/

The Internet Scout Project is designed to show you the best resources on the Internet. Librarians and educators filter the information for you. The report is delivered to your email in-box three times a week. The web site also provides archival copies of previous issues. See Figure 2-5.

EduPage Archives
http://webserv/educom.edu/edupage/edupage.html

Published three times a week and delivered via email, EduPage provides a digest of current events related to technology in education.
To subscribe, send an email message to:
>   listproc@educause.unc.edu

In the message, type:
>   subscribe edupage *yournamehere*
>   (Of course, replace yournamehere with your name.)

To unsubscribe, send an email message to:
>   listproc@educause.unc.edu

In the message, type:
>   unsubscribe edupage

E-conferences

E-conference is an umbrella term that includes discussion lists, interest groups, e-journals, e-newsletters, Usenet newsgroups, forums, and chat groups. These resources monitor a variety of lists. Locate a list on a topic of personal interest and subscribe. Messages posted to the list will be sent to you via email. One caution: given the volume of mail in the e-conference, these tools will certainly get you connected, but, your email in-box will increase dramatically.

The Directory of Scholarly and Professional E-Conferences
    http://www.n2h2.com/KOVACS
AERA ListServs
    http://aera.net/resource/listserv.html
Liszt
    http://www.liszt.com/
The List of Publicly Accessible Mailing Lists
    http://www.neosoft.com/internet/paml/index.html
The List of Lists
    http://catalog.com/vivian/interest-group-search.html
EdWeb's Email Discussion Lists and Electronic Journals
    http://edweb.cnidr.org/lists.html
Reference.COM
    http://www.reference.com

Tactic 2-2   Utilizing news services

The web offers a variety of tools and resources for monitoring current events. This tactic examines multiple ways in which you might receive your daily briefing of world events.

Online news services

USA Today
    http://www.usatoday.com/
The New York Times **on the Web**
    http://www.nytimes.com/
The Washington Post
    http://www.washingtonpost.com
CNN Interactive
    http://www.cnn.com

Sites with collections of daily news

Editor and Publisher Magazine
http://www.mediainfo.com/ephome/npaper/nphtm/online.htm
Access Editor and Publisher's Directory of the World's Online Newspapers. More than 2,560 newspapers are currently available.

NewsLink
http://www.newslink.org/newslink/
Links to 690 newspapers, magazines, broadcast outlets and other related sites. Search by name, region, and other categories.

Newsworks
http://www.newsworks.com
A megasite, compiling more than 125 national, state, and local newspapers.

Newbot
http://www.hotwired.com/newbot/
Scans hundreds of online news sources and web editions of newspapers for keyword, company, or subject.

NewsReal Industry Watch
http://cnnfn.news-real.com
Provides daily coverage of hundreds of stories in 20 categories collected from various wire services and newspapers. Users select a category and then move their mouse over a story to receive an executive summary. One click brings up the full text of the story. An excellent format for scanning the news.

The Electronic Newstand
http://www.enews.com
Described as "The Ultimate Magazine Service," this site allows users to search for magazine subscriptions and explore links to magazine web sites and selected articles.

News monitors

Several services monitor news developments and provide analysis of the developments.

TheNet
    http://www.thenet-usa.com
Clarinet Communications
    http://www.clarinet.com
BusinessWire
    http://www.businesswire.com
NewsPage
    http://www.newspage.com
EarthAlert: Daily Updates on the State of the Planet
    http://www.discovery.com/news/earthalert/earthalert.html

Personalized front pages

One way to deal with overwhelming amounts of information is to use personalized news services. Users complete a profile indicating their interests in various areas (news, stock, premium content, weather, weblinks, websearch) and receive an electronic edition of a newspaper that includes only the articles that match their interest profile.

Crayon
    http://www.crayon.net
InfoSeek
    http://Yournews.infoseek.com/news
Earthlink
    http://www.earthlink.net
MercuryMail
    http://www.mercurymail.com
MSNBC
    http://www.msnbc.com
NewsTracker
    http://nt.excite.com

Content push

The following two sites offer a service in which information is sent or "pushed" to your computer in a continuous stream. Subscribing to an information push service enables you to continuously monitor news, financial, and sports information. Streams can be configured to appear at the bottom of your screen or displayed whenever your screensaver is activated.

Pointcast
http://www.pointcast.com
> Gathers news and information from the web and streams it across your screen. Up-to-the-minute. No subscription fees.

Backweb
http://www.backweb.com
> Select information you are interested in monitoring. Searches and streams the information to your screen--downloads during idle times.

Audio and video broadcasts on the web

The connectivity of the web also allows for audio and video broadcasts enabling you to hear and/or see information. Unlike other web sites that primarily distribute text-based information, the following sites transmit audio files (interviews, audio books, radio, news) and video files (interviews, webtv). These applications are not for computers with weak processing power or slow Internet connections!

AudioNet
> http://www.audionet.com

Audio and Video on the Net from the Digital Jock
> http://www.dj.org

Bookradio
> http://www.bookradio.com/

CableNews Network
> http://www.cnn.com/audioselect

Cnet
> http://www.cnet.com/Content/Radio/
> http://www.cnet.com/Content/TV/

Net Radio
> http://www.netradio.net

RealAudio
http://www.realaudio.com
> Download the free RealPlayer so you can take advantage of web sites using RealAudio and RealVideo for interviews, music, tools, tips, and so on.

TimeCast: The Real Audio Guide
http://www.timecast.com/
> A guide to live events on the web as well as sites with RealAudio and RealVideo. Plus more than 500 radio and TV stations broadcasting over the web.

Tactic 2-3 Utilizing tools that monitor

To save users time, new tools have been developed to assist them in monitoring developments on the web. The products feature a variety of creative strategies.

The New 100 Hot Web Sites
http://www.100hot.com
> Web 21 of Palo Alto, California, uses a combination of surveys, logs, and traffic samples to compile its top-100 list.

Ergo Health Systems
http://www.ergohealth.com
> Ergo Break is computer software that counts the number of times you touch the keyboard, the distance you move the mouse, and how often you press the mouse button. At regular intervals that you can set, the program reminds you it is time to take a break. Developed by the Rehabilitation Institute of Chicago and ERGO Health Systems. Runs with Windows. From the main menu select ErgoProducts and Computer Accessories.

Web monitors

A new class of tools has been designed to automatically monitor the web for changes. Essentially, the user designates selected sites that she would like to monitor, and the software then visits these sites on a predetermined schedule and downloads copies to her local server. This type of monitoring is often done to provide an intranet with information for multiple users, thereby saving them time from individually navigating to the various web sites. In addition, because you can download a copy of selected web sites to your hard drive and access the information without going online. Web monitors also serve as an essential backup strategy for anyone who makes presentations using web sites.

Highlights2
http://www.tierra.com

Smart Bookmarks 3.0
   http://www.firstfloor.com
NetAttache Pro
   http://www.tympani.com
SurfSaver
   http://www.surfsaver.com
WebStorm 2.1
   http://sharpeware.com/html/webstorm.html
WebWhacker
   http://www.bluesquirrel.com

Tactic 2-4 Shifting from online to offline browsing

One of the negative factors often associated with electronic text is the fact that the user must be at the computer to read the information. Many authors have noted that a newspaper is much more portable than a computer. However, the tools supporting this tactic illustrate a new class of products that serve as a "media shift."

Audio Highway
http://www.audiohighway.com

Download selections of digital audio and transfer to the ListenUpPlayer for later playing ($299). This device lets you take Internet audio with you wherever you go (consider all the content listed in Tactic 2-2 above!). Holds up to 2 hours of spoken audio. Downloads are free but every 50 minutes of contents comes with 6 minutes of commercials. As audio content on the web expands, the ListenUpPlayer is a very creative solution for making the content of the web portable.

NewsCatcher
http://www.airmedia.com

The Internet's only wireless news network. Use when your modem is not connected. Requires an internet antenna and AirMedia Live software and service. Windows 95 compatible.

Ricochet Wireless Internet Access
http://www.ricochet.net/

A modem that attaches to your laptop computer and transmits at 28.8 kbps, Ricochet is a wireless modem and service that provides a fast, reliable, secure transmission, with access to the Net from anywhere in Metricom's service area (currently San Francisco Bay area, Seattle, Houston, and Washington, D.C.).

CHAPTER 3
Designing the Study

Because a research study's viability depends on the amount of thought and effort devoted to its design, a researcher devotes a considerable amount of time to tasks involving the conceptualization of a research project and the selection of a research design. This chapter examines seven tasks commonly completed in the process of designing a study. Within this framework, 22 tactics are presented for utilizing technology tools to enhance research productivity when designing research studies.

Task #3 Engaging in Preliminary Exploration of a Topic
       Tactic 3-1 Sending email
       Tactic 3-2 Browsing the web
Task #4 Conducting an Extensive Review of the Literature
       Tactic 4-1 Identifying appropriate databases to search
       Tactic 4-2 Browsing web-based publications
       Tactic 4-3 Considering options for document delivery
       Tactic 4-4 Writing the literature review
Task #5 Selecting an Appropriate Methodology
       Tactic 5-1 Selecting a research design
       Tactic 5-2 Locating resources for qualitative research
       Tactic 5-3 Locating resources for quantitative research
       Tactic 5-4 Locating resources for survey research
       Tactic 5-5 Selecting and describing subjects
       Tactic 5-6 Locating assessment instruments
Task #6 Identifying Potential Sources of Funding
       Tactic 6-1 Subscribing to a funding alert service
       Tactic 6-2 Searching for appropriate funding sources
Task #7 Engaging in Collaboration
       Tactic 7-1 Locating an individual
       Tactic 7-2 Using the web as a telephone
       Tactic 7-3 Using video conferencing
       Tactic 7-4 Using audiographics
       Tactic 7-5 Using work group software
Task #8 Preparing Proposals
       Tactic 8-1 Cutting and pasting information
       Tactic 8-2 Using proposal development tools
Task #9 Participating in Human Subjects Review
       Tactic 9-1 Locating procedures, policies, forms

Task #3: Engaging in Preliminary Exploration of a Topic

Curiosity about a topic is the first step associated with building a research agenda. Researchers, new and experienced alike, often explore concepts and resources for an extended period without a specific research question in mind. The purpose of this exploration is to determine who's who in the field, locate seminal or frequently cited works, and determine how widely the literature is scattered. Two tactics are offered to assist in the task of initially exploring a topic.

Tactic 3-1 Sending email

One of the most effective means to discovering new information about a topic is to ask someone we know to provide some pointers to assist us in getting started. For this strategy, email is an excellent tool. Begin by sending a message to someone you know working in the area you wish to explore. Request some advice on names you should look for, as well as seminal journal articles, books, or web sites you should consult. If your contact is willing to engage in extended conversation, outline your research interests and request feedback. One caution: avoid the typical student request of: "Please send me everything you have on ___ ."

A Beginner's Guide to Effective Email
http://www.webfoot.com/advice/email.top.html
Whereas most pages on this topic focus on a technical introduction, this site, shown in Figure 3-1, examines the style and substance in our electronic communications and explores the nuances of electronic text as a communication form. An essential primer for those new to the online environment.

Eudora
http://www.eudora.com/
One problem with email is how to manage the ever-increasing number of messages. Eudora is an email program that offers powerful filtering, sorting, and filing capabilities for managing your email. Available for both Macintosh and Windows. Download a freeware version and see for yourself before you buy the commercial version, *Eudora Pro*.

See also	Tactic 1-4, Participating in the research community

Figure 3-1
A Beginner's Guide to Effective Email
Used with permission of Kaitlin Duck Sherwood.

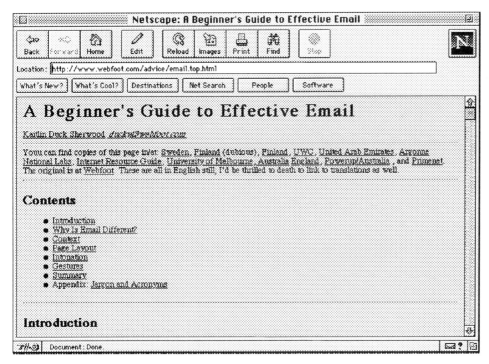

Tactic 3-2 Browsing the web

"Surfing the web" is a wonderful activity as long as you have the time (and patience) to see what turns up. So, of course, set aside time to conduct a net search using one of the many search engines (see Tactic 1-6) and enjoy making serendipitious discoveries. Within this tactic we explore five different strategies for exploring a topic via the web.

See also	**Tactic 1-6, Learn about anything you can imagine**

Authoritative national resources

ERIC Thesaurus Online
http://ericae.net/search.htm

Choose "Search ERIC" then select "ERIC Search Wizard" 2.0 to conduct a search of the ERIC Thesaurus. This is a useful site to visit in order to identify the standardized language used as keywords in most databases indexing materials.

Outline of the Library of Congress Classification
http://lcweb.loc.gov/catdir/cpso/lcco/lcco.html

The authoritative list of classification categories used by the Library of Congress and most academic libraries. Features 21 broad classes that are divided by subclass. Click on the class name to review the detailed classification system for that class. Understanding the structure of knowledge within a discipline may suggest a variety of search strategies.

A National Network of Regional Educational Laboratories
http://www.nwrel.org/national/

Provides links to all of the U.S. Department of Education Regional Educational Labs as well as the ERIC Clearinghouses.

Start with a professional association and explore links

American Education Research Association (AERA)
   http://aera.net
American Psychological Association (APA)
   http://www.apa.org
Association for Computing Machinery (ACM) Special Interest Groups: Computer Uses in Education (SIGCSE)
   http://www.acm.org/sigcue/
Association for Supervision and Curriculum Development (ASCD)
   http://www.ascd.org
Association for Educational Communications and Technology (AECT)
   http://www.aect.org
Council for Exceptional Children (CEC)
   http://www.cec.sped.org
International Society for Technology in Education (ISTE)
   http://www.iste.org
The National Council for Research and Planning (NCRP)
   http://www.raritanval.edu/ncrp
The National Diffusion Network (NDN)
   http://www.ndn.org
National Education Association (NEA)
   http://www.nea.org

National Staff Development Council (NSDC)
   http://www.nsdc.org
Phi Delta Kappa (PDK)
   http://www.pdkintl.org
Society for Computer Simulation (SCS)
   http://www.scs.org

Start with a specialized organization and explore links

Internet Resources for Institutional Research
http://highered.org/links/linkmap.html

A comprehensive site with over 2,400 links in more than 80 categories for students and faculty interested in institutional research. Developed for the Association for Institutional Research and maintained by Professor John Milam, George Mason University.

Figure 3-2
Internet Resources for Institutional Research
Used with permission of John Milam.

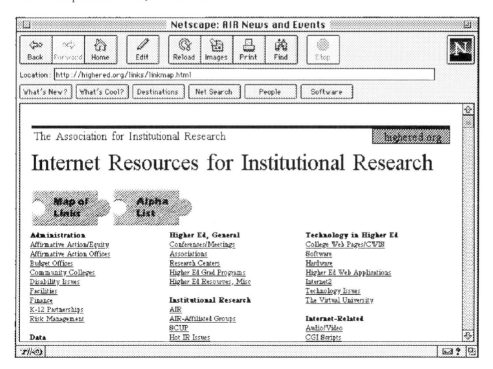

Institute for Research on Learning
http://www.irl.org/pubs/

A research institute at Northwestern University describes its research work on learning and the publications available.

RAND
http://www.rand.org/

A nonprofit institution focusing on the use of research and analysis to improve public policy.

National Clearinghouse for Educational Facilities
http://www.edfacilties.org

Provides information and technical assistance for all facets of school facility planning, design, construction, and maintenance.

gender Inn: Women's and Gender Studies Database
http://www.uni-koeln.de/phil-fak/englisch/datenbank/e_index.htm

A bibliographic database cataloguing more than 5,000 works in feminist theory, feminist literary criticism, and gender studies. The focus is primarily on English and American literature from 1950 to the present.

Perseus Project
http://www.perseus.tufts.edu

Perseus is a digital library of resources for studying the ancient world. The library includes ancient texts and translations, philosophical tools, maps, secondary essays, and extensively illustrated art catalogues. Created from the contributions of more than 70 institutions under the direction of Editor-in-Chief, Gregory Crane, Tuft University.

Social Science Information Gateway
http://sosig.esrc.bris.ac.uk/

SOSIG is an online catalogue of high-quality Internet resources. The catalogue offers users the chance to read descriptions and then access these resources directly. A librarian or academic selects and describes each resource, making this site "the Internet equivalent of an academic research library for the social sciences."

Argus Clearinghouse
http://www.clearinghouse.net

Scouring the Internet for resource guides, Argus Associates (Figure 3-3) describes and evaluates resource guides and makes them available through the clearinghouse. A well-designed site.

Figure 3-3
Home Page for The Argus Clearinghouse
Used with permission of Argus Associates, Inc.

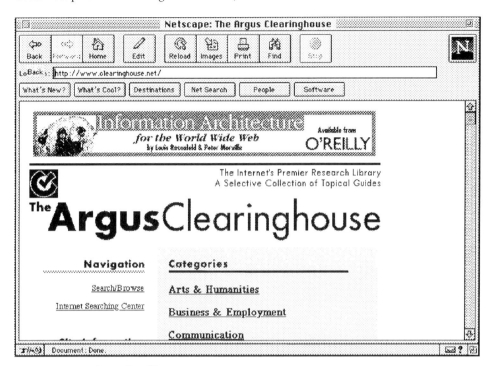

The Idealist
http://www.idealist.org
> A project sponsored by Action Without Borders, this site provides links to 14,000 organizations. Offers extensive resource listings and the opportunity to search by topic, organization, or country.

The Internet Nonprofit Center
http://www.nonprofits.org
> Features a database for locating nonprofit organizations in the United States as well as a variety of information resources.

National Center for Charitable Statistics
http://nccs.urban.org
> The National Center for Charitable Statistics (NCCS) is the national repository of data on the nonprofit sector in the United States.

Meta-Index of Non-Profits
http://www.philanthropy-journal.org/plhome/plmeta.htm
> Features a meta-index of nonprofit organizations and their activities. Sponsored by the *Philanthropy Journal Online.*

Search for information in non-traditional sources

Deja News Research Service
http://www.dejanews.com/

Usenet groups are online topical discussion groups. Although reviewing the thread of a conversation may be tedious, it may also provide important insights about a topic that has yet to be published or widely disseminated. A database indexing the archived Usenet news of almost everything posted in Usenet since early 1995.

National Information Services Corp.
http://www.nisc.com

The National Information Services Corporation publishes a full range of full-text and bibliographic databases. A CDROM (Windows) of particular interest to researchers seeking to locate information found in non-traditional sources is the *The Alternative Press Index*. This index catalogues 250 alternative, radical, and left publications in the United States, 1991 to the present. Quarterly updates.

Search for information in non-text formats

Broadcasting
http://www.lib.umd.edu/UMCP/LAB/biblio.html

This site contains descriptions of audio, film and video, books, photos and transcripts from the collection of American Broadcasting history at the University of Maryland at College Park.

Television News Archive
http://tvnews.vanderbilt.edu/

The archive began taping the evening news broadcasts of the three major networks (ABC, CBS, and NBC) on August 5, 1968. Each newscast has been abstracted and indexed. Search the archive of abstracts of TV news and order videotape copies for research or instruction. Maintained by Vanderbilt University.

The Ultimate TV Show List
http://www.UltimateTV.com/UTVL/

An extensive collection of links to web sites devoted to TV shows. Search or browse more than 10,000 links to some 1,300 TV shows.

Task #4: Conducting an Extensive Review of the Literature

Basic strategies for conducting a literature review are commonly described in most research methods textbooks and courses. Indeed, it is the foundation of the research study. Although students may be inclined to take shortcuts, experienced researchers know that this task is really an investment in the success of your work. The five tactics described below utilize the power of the Internet and a vast array of databases to identify relevant literature and ideas that will establish the need for your research, enhance your research methodology, and provide a conceptual foundation for designing your study.

Tactic 4-1 Identifying appropriate databases to search

Researchers have many choices to consider when selecting appropriate databases to search as part of their review of the literature. This section describes a variety of information sources for indexed literature as well as publications that are more obscure and part of the "fugitive" literature.

Alexa
http://www.alexa.com

Visit this site and download a free copy of Alexa. When installed on your computer, this software adds a toolbar to your Internet browser. Then, as you browse, Alexa provides in-depth information about every site you visit and offers a continuous source of related sites to assist you in finding sites with similar information. An incredible tool!

Global Information Locator Service (GILS)
http://info.er.usgs.gov/gils/index.html

A unique site, GILS is designed to make it easy to locate information of all kinds, in all media, in all languages. Utilizes an international standard for information searching.

WebCATS: Library Catalogs on the Web
http://library.usask.ca/hywebcat/libtype.html

This site allows users to search libraries around the world. Maintained by Peter Scott and Doug Macdonald of the University of Saskatchewan Library.

Internet Accessible Libraries
http://www.ed.gov/EdRes/EdLibs.html
A list of libraries in the United States that are available on the Internet.

Infomine: Scholarly Internet Resource Collections
http://lib-www.ucr.edu/
Maintained by the University of California, this site serves as an introduction to the use of Internet resources for faculty, students, and research staff in higher education. An extremely friendly site that catalogues more than 11,000 information resources.

ERIC
http://www.aspensys.com/eric/
Search the ERIC database using one of several search engines. Gateway to the complete ERIC Clearinghouse system.

National Center for Educational Statistics
http://nces.ed.gov
The mecca of U.S. statistics and information on the condition of education. This site offers access to data sets, analysis tools, publications, and incredible resources of interest to educational researchers.

National Library of Education
http://www.ed.gov/NLE/
The National Library of Education is intended to serve the U.S. Department of Education, its contractors and grantees, the Congress, the Executive Office of the President, education and library associations, and the public. It aspires to be the one-stop information and referral source on U.S. education.

Electric Library
http://www.education.elibrary.com
A comprehensive search tool for researchers of all ages. Free 30-day trial period after you register. Provides extensive access to full-text resources.

Tactic 4-2 Browsing web-based publications

Because of the increased availability of electronic information, the web can be a valuable source of information for the researcher. In contrast to Tactic 4-1, which seeks to examine traditional publication outlets for research results, this tactic illustrates the need to conduct a search of electronic journals as an alternative form of published research.

Scholarly Journals Distributed via the World Wide Web
http://info.lib.uh.edu/wj/webjour.htm

Search a directory of web-based scholarly journals maintained by the University of Houston libraries. An excellent resource.

The Electronic Journals Collection
http://ejournals.cic.net/

An academic consortium of the members of the Big Ten athletic conference and the University of Chicago, this site is a prototype electronic journal management system. By incorporating all freely distributed scholarly electronic journals available online, this site seeks to be an authoritative source of electronic research and academic serial publications.

New Jour
http://gort.ucsd.edu:80/newjour/

The Internet list for journals and newsletters available on the internet. Download the list of materials indexed or navigate an alphabetical list of some 5,800 items. A comprehensive site to say the least!

The On-Line Books Page
http://www.cs.cmu.edu/books.html

Visit this site to read full-text versions of books that are out of copyright. Maintained by John Mark Ockerbloon in cooperation with Carnegie–Mellon University.

Ziff-Davis Publications
http://www.zdnet.com

Looking for information on the technology industry? Visit this site and click on ZD Magazines to search across publications produced by Ziff-Davis. A mega site!

The Conditions of Education, 1997
http://nces.ed.gov/pubs/ce/index.html

A comprehensive annual report on the status of education in the United States. The full text of this report is available along with complete charts of the data. Released annually.

1996 Digest of Education Statistics
http://nces.ed.gov/pubs/D96/

The Digest provides a compilation of statistical information about American education, kindergarten through graduate school. Includes more than 400 tables of data collected from government and private sources. Released annually.

Tactic 4-3 Considering options for document delivery

The traditional form of document delivery involves completing an interlibrary loan request at the libary and subsequently waiting for weeks or months for the book or article reprint you requested. No wonder scholars have sought alternative methods for retrieving information! The resources outlined below offer a variety of options for you to receive information via traditional mail, email, or fax.

Locating a specific bookseller

BookWeb
http://www.ambook.org
Sponsored by the American Booksellers Association, BookWeb is a complete source for information about books, bookstores, authors, and the book industry. Features the CIBON Bookstore Directory to search for a list of more than 4,000 independent bookstores.

The Association of American University Presses
http://aaup.princeton.edu
Learn more about this professional association and search the online catalogue. Lists more than 75,000 book titles from more than 60 university presses.

Ordering books

Amazon.Com
http://amazon.com
Promoted as the world's largest bookstore, this site enables you to order books for delivery to your home.

Asces
http://www.acses.com/
Search this site for a book of interest by title, author, keyword, or International Standard Book Number (ISBN). Then, the web agent will check more than 25 retail online bookstores for availability, price, and shipping time line.

National chains

Barnes and Noble
   http://www.barnesandnoble.com
Borders
   http://www.borders.com
Half Price Books
   http://www.halpricebooks.com

Independent book sellers online

A Clean Well-Lighted Place for Books (San Franciso, CA)
   http://www.bookstore.com
Booksmith (San Franciso, CA)
   http://www.booksmith.com
Cody's (Berkeley, CA)
   http://www.codysbooks.com
Davis-Kidd Booksellers (Nashville, TN)
   http://www.daviskidd.com
Elliott Bay Book Company (Seattle, WA)
   http://www.elliottbaybook.com/ebbco/
Hawley-Cooke (Louisville, KY)
   http://www.hawley-cooke.com
Kepler's (Menlo Park, CA)
   http://www.keplers.com
Midnight Special (Santa Monica, CA)
   http://www2.msbooks.com/msbooks/
Powell's (Portland, OR)
   http://www.powells.com
Tattered Cover (Denver, CO)
   http://www.tatteredcover.com
WordsWorth Books (Cambridge, MA)
   http://www.wordsworth.com

Locating out-of-print books

East Bay Book Company
   http://www.eastbaybooks.com/

Academic publishers: Higher education

Allyn and Bacon
   http://www.abacon.com
Harcourt Brace
   http://www.harcourtbrace.com
Heinemann
   http://www.heinemann.com
JAI Press
   http://www.jaipress.com
Jossey-Bass
   http://www.josseybass.com
Kluwer Academic Publishers
   http://www.wkap.com
Lawrence Erlbaum Associates
   http://www.erlbaum.com
McGraw-Hill
   http://www.mhcollege.com
Merrill
   http://www.merrilleducation.com
Prentice Hall
   http://www.prenhall.com
Macmillan Publishing
   http://www.mcp.com
Research Press
   http://www.researchpress.com
Simon and Schuster
   http://www.simonandschuster.com

Figure 3-4
Home Page for UnCover
Used with permission of CARL Corporation.

Document delivery services

UnCover
http://uncweb.carl.org/

Search the unCover database, shown in Figure 3-4, of nearly 17,000 periodicals using a keyword indexing search. After locating appropriate articles, order copies to be delivered to you via fax, email, or mail from CARL.

ERIC Documents in Electronic Format
http://edrs.com

ERIC is the foremost database of educational documents. However, until recently, you needed to go to the library to retrieve the documents off microfiche. Now, selected ERIC documents are available full-text online from the ERIC Document Reproduction Service (EDRS).

ProQuest Dissertation Abstracts on the Web
http://wwwlib.umi.com/dissertations

Provides access to University Microfilm's Dissertation Abstracts. Free guest privileges allow you to search citations and abstracts for all new titles arriving in the past three months. Before you leave this site, explore UMI to review its family of database products.

Tactic 4-4 Writing the literature review

Writing the literature review is another traditional task familiar to the researcher. Clearly, the tool of choice for this tactic is the word processor. Record your reading notes in one or more files using your word processor. As you begin to compose your literature review by synthesizing the information you have collected, copy and paste within and between documents.

See also	**Tactic 15-1, Ready reference resources for writers**

Task #5: Selecting an Appropriate Methodology

Designing the methodology of a research study is the science component of a researcher's work. Researchers spend a considerable amount of time evaluating possible designs, outlining strategies for selecting their sample, and defining the data collection and analysis procedures. Six tactics are described here to assist in the process.

Tactic 5-1 Selecting a research design

Beginning researchers commonly make errors in selecting the appropriate research design because it requires a fair amount of knowledge. The following products are designed to assist in clarifying assumptions about various research designs.

Methodologist's Toolchest
http://www.sagepub.com/sagepage/mtc.htm
A CDROM-based product for Windows that helps researchers in multiple areas of their work: proposal development, research design, sample size, and data collection. Features modules: Peer Review Emulator, Statistical Navigator, Ex-Sample, Designer Research, WhichGraph, Data Collection Selection, Measurement and Scaling Strategist, ETHX, and HyperStat.

SRM Database of Social Research Methodology
http://www.sagepub.com
A comprehensive database on CDROM of more than 40,000 references to literature on behavioral methodology, statistical analysis, and computer software in social science books and journals. Available in one-year subscriptions of network versions.

The Political Methodology Society
http://wizard.ucr.edu/polmeth/polmeth.html
> This is the home page of the Political Methodology Society, Political Methodology Section of the American Political Science Association. Check out the links in the computing section to locate Internet sites related to software, methodology, and political research.

Theory Construction and Research Methods
http://www.orst.edu/dept/hdfs/hdfs-sub/tcrm.html
> Affiliated with the National Council on Family Relations, this home page describes the workshops available on theory construction and research methods.

The Sin of Omission–Punishable by Death to Internal Validity: An Argument for Integration of Qualitative and Quantitative Research Methods to Strengthen Internal Validity
http://trochim.human.cornell.edu/gallery/bowen/hss691.htm
> This page is the fulltext of a document by Kathryn A. Bowen, Cornell University. Provides a technical analysis that compares and contrasts qualitative and quantitative research methods.

Tactic 5-2 Locating resources for qualitative research

Researchers interested in qualitative research methods will find an increasing amount of information and software tools online. The following tools will assist researchers in connecting with others that share an interest in qualitative research methods.

See also	**Tactic 1-1, Identifying resources for learning about research** **Tactic 1-2, Identifying resources for teaching about research** **Tactic 14-1, Using generic tools for text analysis** **Tactic 14-2, Using specialized text analysis tools**

The Qualitative Research Page
http://www.irn.pdx.edu/~kerlinb/qualresearch/qualPage.html
> A well-designed site that introduces topics and resources of interest to qualitative researchers and students.

Figure 3-5
QualPage, Resources for Qualitative Researchers
Used with permission of Judy Norris.

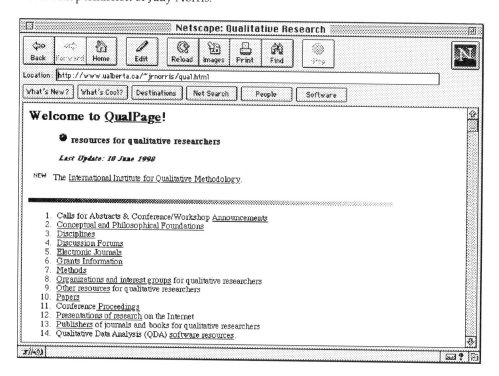

QualPage
http://www.ualberta.ca/~jrnorris/qual.html
Developed and maintained by Judy Norris, this site is a comprehensive index to qualitative research resources. Its clean design enables fast loading and its coverage is distinctively international and multidiscipinary. See Figure 3-5.

Comparing Qualitative and Quantitative Methods
http://trochim.human.cornell.edu/kb/qual.htm
A thoughtful overview about key issues associated with selecting a methodology. Developed by William M.K. Trochim of Cornell University. Follow the link back to the KnowledgeBase home page and discover Professor Trochim's *Online Research Methods Textbook.*

Computer Software for Research
http://www.irn.pdx.edu/~kerlinb/researchware.html
Developed and maintained by Dr. Bobbi Kerlin, Portland State University. Provides a series of links to downloadable software for qualitative and quantitative analysis.

Qualitative Solutions and Research
http://www.qsr.com.au/

This is the home page for the company that developed QSR NUD•IST qualitative software. This software is a comprehensive system for the development, support, and management of qualitative data analysis projects. It is ideal for processing non-numerical unstructured data with powerful tools for indexing, searching, and theory-building.

Computer Assisted Qualitative Data Analysis
Software Network Project
http://www.soc.surrey.ac.uk/caqdas/website.htm

This site describes the work of a project funded by the United Kingdom Economic and Social Research Council, whose purpose is to describe and disseminate the practical skills needed to use software designed to assist qualitative data analysis (e.g., field research, ethnography, text analysis). Offers information on useful resources such as short courses, downloadable software, and a bibliography.

The Ethnograph Web Site
http://www.qualisresearch.com/

The is the home page for Qualis Research Associates, developers of Ethnograph software. Provides extensive product information and support.

The Multimedia Ethnographic Research Lab
http://www.merlin.ubc.ca/

This site describes the work of the Faculty of Education at the University of British Columbia to utilize multimedia tools to conduct enthnographic research. Established in 1991, this R and D lab seeks to develop theories and tools for conducting ethnographic inquiry using emerging technologies.

Museums Online
http://www.okc.com/morr/wwwCategories/WWWEthnography.
html

One way to explore online museums is to examine their collections using different themes. This site offers a categorical listing of online museums and their collections that involve ethnography.

Tactic 5-3 Locating resources for quantitative research

Quantitative researchers will find an extensive amount of information online concerning research methods, tools, and data sets. This section provides links to quantitative research methods.

See also	Tactic 1-1, Identifying resources for learning about research
	Tactic 1-2, Identifying resources for teaching about research
	Tactic 13-1, Using generic tools for data analysis
	Tactic 13-2, Using specialized data analysis tools

Statistical Archives
http://www.uvm.edu/~dhowell/StatPages/Archives.html

A comprehensive archive to statistical information developed and maintained by David C. Howell, University of Vermont. One especially useful feature is the option of searching for questions and answers in the archives of various statistical online discussion groups. His bookmark collection is exemplary.

The Globally Accessible Statistical Procedures (GASP) Page
http://www.stat.scarolina.edu/rsrch/gasp/

A collection of online statistical procedures, that you can access from your web browser. These are actual routines that take data (entered via a form) and return statistical results. Includes some CGI and Java routines.

A Compilation of Statistical Material
http://www.lib.umich.edu/chdocs/statistics/stat_guide_home.html

A compendium of statistical resources available via the web. Includes links to data sets, FTP sites, mailing lists, and much more. Assembled by Ken Varnum and John Weise at The University of Michigan.

Computer Software for Research
http://www.irn.pdx.edu/~kerlinb/researchware.html

Developed and maintained by Dr. Bobbi Kerlin, Portland State University. Provides a series of links to downloadable software for qualitative and quantitative analysis.

Figure 3-6
Y? The National Forum on People's Differences Home Page
Used with permission of Phillip J. Milano.

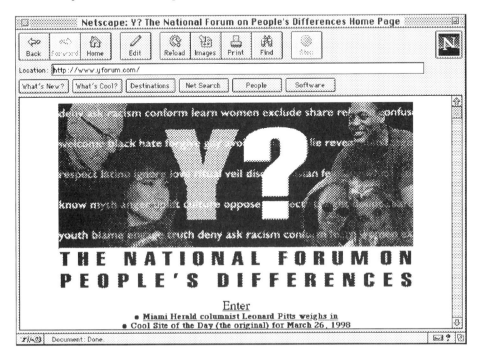

Tactic 5-4 Locating resources for survey research

Survey research often involves questionnaires and interviews. A variety of tools have been developed to facilitate the process of constructing appropriate surveys as well as aid in the process of data collection and analysis.

Y? The National Forum on People's Differences
http://www.yforum.com/

An online environment designed to provide visitors with a means to ask people from other ethnic or culture backgrounds questions they've been too uncomfortable to ask. Topics include those related to differences in age, class, disability, gender, geography, occupation, race/ethnicity, religion, and sexual orientation. Clear guidelines for asking and answering questions are provided to enable participants to discuss sensitive topics without fear of attack. See Figure 3-6.

Perception Analyzer
http://www.cinfo.com

A system that includes hardware and software, the Perception Analyzer is an example of a electronic group measurement system or audience response system. It was originally designed for improving communication in focus groups. Each participant's hand-held dial relays his or her response via radio frequency to an interface connected to a PC. The installed software assembles the data according to the user's specifications and displays the results on an external monitor or screen. The Perception Analyzer has been used extensively in academic environments and meeting facilitation, as well as in marketing research, public opinion research, media research, and jury research.

Survey Select
http://www.surveyselect.com/

A CDROM-based product for Windows assists users in designing, administering, and analyzing surveys. Also includes features to create web-based survey instruments.

Decisive Survey and Decisive Feedback
http://www.decisive.com

Decisive Survey builds a questionnaire and distributes it to the email addresses you specifiy. It then tracks, gathers, tabulates, and charts the responses easily and quickly. It will also build a web survey and email you back the responses. Download a free 30-day trial of Decisive Survey (a Windows-based product). Decisive Feedback is a web-based product that facilitates the process of getting feedback from your web page visitors.

SurveyWin
http://www.raosoft.com/

Database software for designing, collecting, and analyzing survey data. Developed by Raosoft, Inc., these tools support surveys distributed via disks, LANs, or the web.

MaCATI 2.5
http://www.senecio.com

Helps users automate survey research tasks including creating and editing questionnaires, and collecting and analyzing data. Provides tools for conducting telephone surveys, sending surveys by mail, and polling visitors to a www site. This Macintosh software is leased by the quarter or year.

Survey Research Center, University of Michigan
http://www.isr.umich.edu/src/
A nationally recognized center on conducting survey research. This site describes the current projects of the center and provides information on its annual summer institute.

Opinion Research Corp.
http://www.opinionresearch.com
The home page of the international organization, Opinion Research Corporation, a premier provider of market research.

National Decisions Systems
http://www.ends.com
A comprehensive provider of marketing research services and products. Extensive descriptions of the software products and sample reports provide a useful overview of how research can inform decision-making.

Market Statistics
http://www.marketstats.com
Market Statistics provides demographic and economic data solutions for firms in a variety of industries. Explore the products and services described at this site. Of particular interest to researchers are the geomarketing systems which merge marketing data with geographic information systems. Also of interest is the Buying Power On-Line system which merges marketing data with Internet delivery of custom reports, and maps of demographic and consumer economics.

Cyber Dialogue
http://www.cyberdialogue.com/frame.html
Cyber Dialogue is an online market research company. Its premier product by the same name uses online research and data mining technologies to enable users to understand the needs and interests of visitors to their web site.

USA Data
http://usadata.com
Collects and distributes data on marketing, company, advertising, and consumer behavior. Specific research categories include demographics, Internet, computers, cellular telephone, retail shopping, travel, finance, automotive, sports promotion, and lifestyles.

Marketing Tools
http://www.marketingtools.com

This site is sponsored by *American Demographics* magazine and serves as a search engine to products to enhance marketing research. Povides links to full-text articles, print publications, and online resources.

Tetrad Computer Applications
http://www.tetrad.com

This site features several software products: MapInfo, a tool for placing data on maps; PCensus, a tool to retrieve demographics for user-defined, standard census and postal areas; and PSearch, used to find target areas with households that match user-defined demographic profiles. Extensive descriptions and screen prints provide a clear indication of the power and flexibility of each tool. Download free demo copies.

CoStar
http://www.costar.com

LabelWriter II is a mailing label printer for Macintosh and Windows computers that supports design and printing of labels of all sizes. An essential tool for conducting mail surveys.

Tactic 5-5 Selecting and describing subjects

It is is important to select the optimum number of subjects for a study to account for attrition and ensure power is adequate to utilize specific analysis procedures. A variety of tools are available to assist the researcher in selecting an appropriate sample.

SamplePower
http://www.spss.com/spower/

A CDROM for Windows95 to help researchers determine appropriate sample size.

Methodologist's Toolchest
http://www.sagepub.com/sagepage/mtc.htm

A CDROM-based product for Windows that helps researchers in multiple areas of their work: proposal development, research design, sample size, and data collection. Features modules: Peer Review Emulator, Statistical Navigator, Ex-Sample, Designer Research, WhichGraph, Data Collection Selection, Measurement and Scaling Strategist, ETHX, and HyperStat. Download a free demo.

nQuery Advisor
http://www.StatsolUSA.Com
A software product for Windows 3.1 or higher, provides computational assistance of sample size and power detemination problems.

American Demographics
http://www.demographics.com
This site is the online companion of the journal by the same name. Presents practical applications of demographic studies as well as technical issues involved in sampling.

Census Counts 98
http://www.decisionmark.com
A software product that provides complete census data, including estimates from 1997 and projections for 2002. Invaluable for creating research samples that are representative of national or local populations.

State Department of Education Agencies
http://www.ed.gov/NCES/sites5.html
Looking to find information on educational issues in a particular state? This site provides links to each state department's education agency. From there you can navigate a variety of educational research and practice issues.

QED
http://www.qeddata.com
A company that focuses on school marketing. QED devloped and maintains the National Education database, Project EdTech, and the Teacher Registry. Issues periodic research reports and offers a broad range of products and services.

Market Data Retrieval
http://www.marketdata.com
A marketing research company that produces mailing list and market reports on trends in K-12 education based on annual surveys of schools.

American School Directory
http://www.asd.com/
The Internet Guide to all 106,000 K-12 schools in the United States. Search for an individual school or for all schools within a state, county, or postal code.

The School Report
http://www.nsrs.com

This free site is designed to provide information about schools when a family is relocating. Uses a multistep search process: state, and city or county. Provides comparative information on schools (size, number of teachers, average class size, etc.). Offers links to school maps, child care, and real estate professionals.

HotList of K-12 Internet School Sites
http://www.gsn.org/hotlist/index.html

Access the Internet sites of all K-12 schools, alphabetized by state. Of course, this list does not identify all schools because each school has yet to develop an Internet presence.

Tactic 5-6   Locating assessment instruments

Norm-referenced assessment instruments are often used in research studies. Several resources are available to assist researchers in locating an instrument or to obtain information about the technical adequacy of a specific instrument.

The ERIC Clearinghouse on Assessment and Evaluation
http://ericae.net

A comprehensive site concerning assessment and evaluation, provides links to other sites involved in educational research, assessment, and evaluation. The ERIC/AE Test Locator, co-sponsored by The Educational Testing Service (ETS), Buros Institute of Mental Measurements, and Pro-Ed test publishers, provides a search engine with links to 10,000 assessment instruments and their availability, as well as texts where the instruments have been reviewed. Be sure to check the section on Partners Program for Professors of Educational Research.

National Center for Fair and Open Testing
http://www.fairtest.org

A nonprofit advocacy group concerned about the misuse of standardized testing in education.

Figure 3-7
The ERIC Clearinghouse on Assessment and Evaluation Web Site

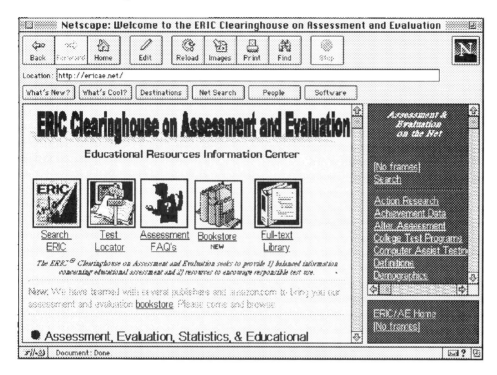

Task #6: Identifying Potential Sources of Funding

Although there is no shortage of good ideas, there are, unfortunately, limited funds to support research. Researchers may spend a considerable amount of time searching for appropriate funding agencies and preparing proposals. Successfully matching your research agenda with the priorities of a funding agency means you'll have the funding needed to implement your ideas. In this section we examine two tactics for facilitating the process of obtaining research funding.

Tactic 6-1 Subscribing to a funding alert service

Subscribing to a funding alert service is a simple but powerful means to utilize technology to assist in identifying suitable funding agencies. Typically, this process involves completing a profile regarding your research interests and indicating the types of funding agencies (e.g., private, government, corporate) you would like to have monitored. Then, whenever the database is updated your profile is run to see if there are any matches. You are subsequently notified by email concerning the new opportunities that appear to match your research interests.

Community of Science Funding Opportunities
http://www.cos.com

After you've added your expertise profile to COS, the COS Funding Alert will automatically match your research interests with information drawn from the COS Funding Opportunities database.

Fedixopportunity Alert
http://nscp.fie.com/

Email announcements of grant opportunities from 12 U.S. federal agencies within 1 to 2 days following each agency's announcements.

IRIS
http://www.LIBRARY.uiuc.edu/default.htm

Illinois Researcher Information Service allows users to create their own search profiles and then receive information on more than 6,900 funding opportunities. Alerts are delivered to user's email account according to time line specified (weekly, monthly, bimonthly).

Eureka Alert
http://www.eurekalert.org

Site sponsored by the American Association for the Advancement of Science. Register for news about scientific books, grants, awards, and meetings. Links to science visuals and reference works. Free.

Tactic 6-2 Searching for appropriate funding sources

The utility of the web as a tool to search for funding is unmatched. The following tools illustrate the variety of resources that are available to assist you in the process of identifying appropriate funding sources to support your research.

A Grant Seeker's Guide to the Internet: Revised and Revisited
http://www.researchgrant.com/IART.HTML

A classic guide that has been revised and updated by Research Grant Guides, Inc. This document provides a detailed how-to introduction to using the Internet to locate information about grant opportunities and how to obtain grant funding.

The Foundation Center's User-Friendly Guide
to Funding Research and Resources
http://fdncenter.org/onlib/ufgtoc.html

An excellent guide for grant writers. Provides valuable information on all parts of the grant development and submission process.

Know About Grants
http://www.ed.gov/pubs/KnowAbtGrants

Electronic text of a practical and helpful booklet, "What Should I Know About ED Grants?" A cradle-to-grave overview of the complete discretionary award grant cycle. Helpful information in applying or administering a grant award from the United States Department of Education.

Federal Register-U.S. Department of Education Documents
http://www.ed.gov/legislation/FedRegister/announcements/index.html

This site contains announcements made in the Federal Register involving Department of Education activities. Includes materials from October 1995 to present. This site is especially useful for locating announcements of new grant competitions.

Federal Register
http://www.access.gpo.gov/su_docs/aces/aces140.html/

Search the Federal Register, the official daily publication of the federal government, for rules, proposed rules, and notices of federal agencies and organizations as well as Executive Orders and other Presidential documents.

The Chronicle of Philanthropy
http://philantrophy.com

The online version of the weekly newspaper of the nonprofit world. Reviews issues involving gifts and grants, fund-raising, workshops, and conferences.

The Grantsmanship Center
http://www.tgci.com

The web site of the Grantsmanship Center, a marvelous resource for learning about the grant development process. Features a variety of how-to publications, workshops, and valuable tips on writing and winning grants.

Council on Foundations
http://www.cof.org

A nonprofit membership association of grant-making foundations and corporations. Provides useful links and resources for locating funding from private foundations, corporations, community foundations, and public charities.

The Foundation Center's Home Page
http://fdncenter.org/

A comprehensive site for grant seekers and grant makers. Extensive information and links.

Oryx Grants Home Page
http://www.oryxpress.com/grants.htm
> Oryx publishes a variety of print and electronic resources on locating funding sources, preparing proposals, and managing grants.

The Grants Keyword Thesaurus
http://www.rams-fie.com
> Provides a common vocabulary for researchers to identify and categorize grant opportunities. Available in a downloadable file. Maintained by RAMS-RIE, developers of a family of grant management tools.

Task #7: Engaging in Collaboration

Because research is an increasingly collaborative effort, there is a growing need for the researcher to have access to tools that support the collaborative process. In this section we will examine four tactics for using technology to support collaborative work.

Tactic 7-1 Locating an individual

The efficiency of email for frequent communication among individuals separated by time or distance is hard to beat. Additionally, the option of using attachments to send files, graphics, and so on makes email an essential tool for collaborative work. However, in the process of designing a research proposal it may be necessary to contact someone with whom we have lost contact or someone we've read about in the literature. Several services can assist in locating individuals and their email addresses.

WhoWhere?
> **http://www.whowhere.com/**

Bigfoot
> **http://www.bigfoot.com**

Switchboard
> **http://www.switchboard.com/**

Four11
> **http://www.four11.com**

Internet 800 Directory
> **http://inter800.com**

Internet Address Find
> **http://www.iaf.net**

InfoSpace
   http://www.infospace.com
World Pages
   http://www.worldpages.com

1998 Higher Education Directory
http://www.hepine.com
> A description of the publication, the *Higher Education Directory*, that lists degree-granting, postsecondary institutions with recognized accreditation. Updated annually, a valuable tool for locating degree programs at any university.

World Academic Database on CD-ROM
http://www.macmillan-reference.co.uk
> A multimeda collection that combines the *International Handbook of Universities* and the *World List of Universities* to create an extensive directory of institutions around the world. From the main menu, select "multimedia and CD-ROM." Windows compatible.

Tactic 7-2 Using the web as a telephone

One of the more mind-boggling applications of the web involves using it like a telephone (web telephony) but without the long distance charges. While some special equipment (software, sound capabilities on your computer, and a modem) are needed to implement this tactic, the investment is quickly repaid through the reduction in your long distance telephone charges. That's right: currently this system operates anywhere around the world for the price of a local phone call. Originally, many of these types of programs allowed you to talk only with others using the same software/system. New standards have resulted in product improvements that now allow cross-platform communication as well as phone calls to regular telephones.

Internet Phone
   http://www.vocaltec.com/
CoolTalk
   http://home.netscape.com/comprod/products/navigator/
   version_3.0/communication/cooltalk/index.html
WebPhone
   http://www.netspeak.com
The Phone Zone
   http://www.phonezone.com

Big Picture Video Phone
  http://www.3com.com/client/pcd/products/bigpicture/
  products/videophone.html
ViaTV Phone
  http://www.8x8.com/
Virtual Voice Internet Telephony
  http://www.virtual-voice.com

Hello-Direct
http://www.hello-direct.com

A firm specializing in telephone productivity tools. An excellent source for all types of telephone equipment and accessories.

Voice on the Net Coalition
http://www.von.org

New legislation is frequently proposed to regulate net phones. This site is the home page of an advocacy group devoted to preventing excessive regulation of web telephony.

Tactic 7-3 Using video conferencing

The cost of video conferencing used to be prohibitive to anyone but large multinational companies. However, with the invention of Connectix QuickCam, anyone can conduct a video conference. A great array of prices and services are available in this area of exploding development.

Full-feature commercial video conferencing

Live Share Plus 4.0
  http://www.picturetel.com
Team Station
  http://www.intel.com/proshare/conferencing/demo
Smart Videoconferencing
  http://www.vtel.com/
Digiphone
  http://www.digiphone.com
Auditorium 1.0
  http://www.placeware.com

Low-cost video conferencing

NetMeeting 2.0
http://www.microsoft.com/netmeeting
Free. Rapidly becoming a new standard for video conferencing on the web.

CoolTalk
http://home.netscape.com/comprod/products/navigator/ version_3.0/communication/cooltalk/index.html
A browser plug-in that provides video conferencing capability as well as a whiteboard, email, and more.

Look@Me
http://collaborate.farallon.com/www/look/download.html
A free, real-time Internet collaboration tool that allows users to watch the screen activity taking place by another Look@Me user from within their browser. It's available as a Netscape plug-in with a stand-alone applet for Windows 3.x.

QuickCam
http://www.connectix.com
Available in black and white or color, these cameras plug into the back of your computer and with the software provided you can take still image photos or make QuickTime movies. Of course, once the images are in the computer you can edit them, send them to others via email, and so on.

CU-SeeMe
http://www.wpine.com/Products/CU-SeeMe/
The premier site devoted to the use of free CU-SeeMe software that can be used with Connectix QuickCams to allow video conferencing over the web. Provides extensive links to sites illustrating how CU-SeeMe can be used.

Surf Check
http://www.surfcams.com/~surfcheck
Updated daily, this site provides daily beach photos, video clips, and surf forecasts from 90 beachcams. One example of the many cam sites on the Internet. Here, you'll be able to check the status of the waves before heading out to the beach.

Campus WebCams
http://141.211.140.236/webcam/webcam.htm
A directory of colleges and universities with webcams. Sponsored by the Society for College and University Planning.

Resources

Audio and Video on the Net
http://www.dj.org

This site is maintained by the Digital Jock, who organizes a range of audio and video files and services that illustrate the multimedia nature of the Internet.

Videoconferencing ATLAS
http://www.savie.com

A resource site on the use of video conferencing in education. Features information on getting started, distance education, navigating the market-place, and more.

Tactic 7-4 Using audiographics

Audiographics involves the transmission of voice and graphic images, used historically for distance education. Today, this technology is being applied to create interactive electronic whiteboards. Images from the boards can be sent over the net, saved in a file, printed, or faxed.

SMART Board
http://www.smarttech.com

The SMART Board is an interactive whiteboard that serves as a projection screen for your computer and enables you to mark on the board just like a regular whiteboard. Images may be saved, stored, printed, or transmitted.

The Digital Flipchart
http://www.tegrity.com

Captures notes and diagrams created on a standard whiteboard using a small video camera. Images are saved to a Windows 95 computer for later review or printing. Useful for meetings or classes.

Ibid Whiteboards
http://www.microtouch.com

Software and hardware that attaches a standard whiteboard to a PC, Ibid preserves records of meetings and makes it easy to distribute notes.

Tactic 7-5 Using work group software

The tools described in this tactic illustrate some of the unique possibilities of collaborative work efforts as a result of the convergence of communication and information technologies. Each product offers a variety of tools to support the communication and collaboration effort of groups, including calendars, whiteboards, chat functions, document creating and editing, and accessing reference materials.

LotusNotes, Domino
   http://www.notes.net/
GroupWise 5
   http://www.novell.com/groupwise/index.html
SmartSuite 97
   http:// www.lotus.com
TeamTalk
   http://www.traxsoft.com/traxsoft/
Face to Face
   http://www.crosswise.com
The Virtual Meeting
   http://www.rtz.com
TeamWave Workplace
   http://www.teamwave.com/
AltaVista Forum
   http://altavista.software.digital.com

ICQ
http://www.icq.com

Pronounced, "I seek you," ICQ is an innovative technology tool that combines email and video conferencing. When the software is installed you are assigned a Universal Internet Number (UIN). Users create a personal profile and a contact list. As part of your profile, you can provide as much or as little information as you would like. Also, you select which associates you would like notified when you are online. Then, whenever you are online, ICQ tells you if any of your colleagues are online so that you can conduct a real-time chat.

Groupwork resources

**GROUP '97. Proceeedings of the International
ACM SIGGROUP Conference on Supporting Group
Work: The Integration Challenge
http://www.acm.org/pubs/contents/proceedings/cscw/266838/
index.html**

> The full proceedings of a conference held Nov. 16-19, 1997 in Phoenix,
> Arizona, concerning the support of group work. The full text of more than
> 40 conference papers is available in .pdf format. Topics include collaborative
> tools, virtual classroom and communities, knowledge management, collabo-
> rative filtering and workflow, and process automation.

Task #8: Preparing Proposals

The essential tool for developing a grant proposal is the word processor.
Whether creating a proposal individually or assembling the input of a team of
writers, researchers must have mastery of the basic keyboarding functions (type,
backspace) and formatting functions (i.e., tab, line spacing, font, style, pagina-
tion) and of course, the spell checker.

Tactic 8-1 Cutting and pasting information

To save time and energy, researchers with experience in preparing research
proposals may re-use portions of their previous work. It is important to high-
light the fact that cut and paste works not only within documents, but also
between documents. Try this with your word processor: Open two documents
on the desktop. Select a section of text from one document, copy, use the view
function from the menubar to move to the second document, place your cursor in
the desired spot, and then use the paste command to insert the text from the first
document. This technique is extremely useful when combining text from multiple authors as well as recycling text from previous proposals.

Tactic 8-2 Using proposal development tools

Recognition of the value of technology in proposal development has led to recent
exploratory projects to facilitate proposal development and submission.

**NSF Fastlane
http://www.fastlane.nsf.gov**

> An example of an electronic proposal submission process used as an experi-
> mental program by the National Science Foundation.

GrantWorks
http://www.biostat.wisc.edu/grantworks/

> GrantWorks is a computerized solution to the National Institutes of Health grant applications PHS 398 and 2590, for managing data and records as well as completing the application form pages. Developed by the University of Wisconsin-Madison, Medical School Biomedical Computing Group, this product represents a creative solution to the tasks associated with preparing a grant application. Download a beta copy of the Windows or Macintosh software.

Task #9: Participating in Human Subjects Review

An important consideration in any type of research is the protection of human subjects from any adverse consequences as a result of their participation in the research. Researchers will need to prepare a proposal to have their proposed work reviewed. Usually this is accomplished through an Institutional Review Board (IRB).

Tactic 9-1 Locating procedures, policies, and forms

Human Subjects
http://www.psych.bangor.ac.uk/DeptPsych/Ethics/
HumanResearch.html

> A comprehensive site with links to position papers, ethics committees, institutional review boards, and courses on ethics.

OPRR Human Subject Protection
http://www.nih.gov/grants/oprr/library_human.htm

> Maintained by the NIH Office of Extramural Research, provides full-text of pertinent regulations, interpretative resources, and links to announcements of OPRR workshops dealing with the protection of human subjects.

The Rockefeller University Institutional Review Board
for Human Research
http://clinfo.rockefeller.edu/irb/irb-ind.htm

> Outlines one university's procedures for institutional review of human subject research.

Human Subjects Research Form
http://orsp1.adm.binghamton.edu/Campus_Forms/Online_Forms/online_humansub.cgi
>An example of an online form for review of human subjects research.

Methodologist's Toolchest
http://www.sagepub.com/sagepage/mtc.htm
>A CDROM-based product for Windows that helps researchers in multiple areas of their work: proposal development, research design, sample size, and data collection. Includes a module dealing with ethics. Download a free demo.

CHAPTER 4
Conducting the Study

Once a research project has been developed, and in many cases funded, the focus of energy shifts from planning to actually conducting the study. In this chapter, we will examine three common tasks involved in conducting a study: creating materials, collecting data, and communicating with field sites. Chapter 7, Research Administrator, provides additional ideas and resources for starting a research project.

Task #10 Creating Materials
       Tactic 10-1 Creating print materials
       Tactic 10-2 Using rapid prototype tools
       Tactic 10-3 Creating web-based materials

Task #11 Collecting Data
       Tactic 11-1 Using hand-held data input systems
       Tactic 11-2 Using portable keyboards
       Tactic 11-3 Using computer-based data collection tools
       Tactic 11-4 Using unobtrusive computer-based data collection tools
       Tactic 11-5 Using web-based data collection tools
       Tactic 11-6 Using voice input
       Tactic 11-7 Using digital cameras

Task #12 Communicating with Field Sites
       Tactic 12-1 Using email

Task #10: Creating Materials

To conduct a research study, the research team often must generate a variety of materials ranging from a simple word-processed protocol for conducting an interview to sophisticated designs that integrate text and graphics. Three tactics are described here to assist researchers in locating appropriate tools for creating print materials, developing prototypes, and creating web-based materials.

Tactic 10-1 Creating print materials

The most common tool for creating print materials for a study is the word processor. A single software program will enable the research team to create most of the materials needed to conduct a study.

Although word processors have incorporated many features that were formerly reserved for desktop publishing programs, occasionally users find the need for advanced formatting and design features that are difficult or impossible to complete using the word processor. Mastery of a desktop publishing program will enable the researcher to design high-quality materials. Some common desktop publishing programs include:

QuarkXPress
   http://www.quark.com
PageMaker
   http://www.adobe.com/prodindex/pagemaker/main.html
Microsoft Publisher
   http://www.microsoft.com/products/

Another useful tool for designing instructional materials is screen capture software. Essentially, screen capture software allows you to take a picture of the screen. A basic screen capture feature is built into the operating systems of both the Macintosh (Press Option-Shift-3) and Microsoft Windows (Press PrntScrn) to create a graphic file of the current screen. This technique is invaluable to illustrate what users will see on the screen when designing training materials. Screen shots may be edited in a paint program, pasted into presentation software, or simply used as clip art in your word processor. Commercial products that provide additional options are:

FlashIt!
   http://www.symantec.com
Print Screen Deluxe
   http://www.jesoft.com
SnagIt
   http://www.techsmith.com

Tactic 10-2 Using rapid prototype tools

Research often involves the examination of new theories or practices. Before embarking on full-scale development it is necessary to test out the ideas on a limited scale and provide evidence of the functionality or "proof of concept." Rapid prototype tools enable non-programmers to design computer-based materials to illustrate the "look and feel" of an end-product.

MacroMedia Director
http://www.macromedia.com/software/authorware

The industry strandard for creating interactive instruction. Files that are created are compatible with both Macintosh and Windows. Explore this site to learn more about the rest of the MacroMedia family of products, including Director, Dreamweaver, FreeHand, and Shockwave.

ToolBook II
http://www.asymetrix.com

ToolBook II is a powerful authoring product for creating computer-based training. Lessons may be delivered via CDROM or the web. For Windows or Unix.

Discovery Systems
http://www.discoverysystems.com

Visit this site and explore the usefulness of CourseBuilder, an authoring tool for creating electronic courseware. An extensive collection of tools for designing, delivering, and assessing instruction.

Digital Chisel 3
http://www.pierian.com

A multimedia hypertext program that enables users to create instructional materials. Includes sophisticated routines for branching and data collection. Ideal for creating electronic true/false and multiple-choice tests. This new version produces java code, which means your files are ready for the web.

HyperStudio
http://www.hyperstudio.com

A easy-to-use, easy-to-learn product for creating multimedia projects. Very popular in K-12 schools as a design tool for students. Web-compatible files means it is easy to publish your project on the web.

Form.Z
http://www.formz.com

FormZ is an advanced 3-D modeling technology used to create 3-D models of products, urban environments, and so on. Designed by auto-des-sys, Inc., they seek to bring supercomputing power to the desktop computer.

Adobe PhotoShop
http://www.adobe.com/proindex/photoshop/main.html

The standard photo design and production tool, this product is the professional illustrator's tool of choice. A researcher proficient in PhotoShop can create logos, original illustrations of models, procedures, and so on.

ED Tech Tools
http://motted.hawaii.edu

This site was created by the University of Hawaii's Educational Technology Research and Development Center. Among its many useful resources is the Quiz Center Online Service, a free service that helps educators who don't know HTML create interactive quizzes on the Internet. When a student completes the quiz, tools at the site can automatically correct and email it to the instructor.

Question Mark
http://www.questionmark.com/

Question Mark is an authoring tool that allows users to create computerized quizzes, tests, and surveys without any knowledge of computer programming. The software allows the author to design feedback for students and saves all students' answers for scoring and analysis. Question Mark's Reporter software module assists the author in analyzing the results. Available for PCs, Macintoshes, LAN or the web. Download a free 30-day trial copy.

Electronic Courseware Systems
http://www.ecsmedia.com

This company focuses on music instructional software, but also features instruction and training software for English, social sciences, and AutoCad.

Figure 4-1
Example of a web-based form used by the U.S. Office of Educational
Research and Improvement for collecting data to inform policy on the
issue, "What do we need to know to improve learning?"
Source: http://www.ed.gov/offices/OERI/RschPriority/learning.html

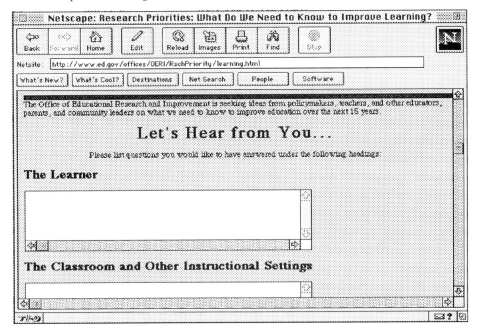

Tactic 10-3 Creating web-based materials

Researchers have begun to explore the use of the web as the instructional delivery system for their research. As a result, they design the content on their web page and research participants connect to the page and interact with the information, see Figure 4-1. Some common tools for creating web pages include:

PageMill
>   http://www.adobe.com/proindex/pagemill/mail.html

FrontPage
>   http://www.microsoft.com/products

Claris HomePage 3.0
>   http://www.clarishomepage.com/products/claris/clarispage/
>   chpbeta_ hdr. html

TASL Training and Seminar Locators
http://ww.tasl.com/tasl/home.html
>   Use this site to find educational, training, and development resources. TASL tracks seminars and allows users to search by event, product, or provider.

Task #11: Collecting Data

Data collection is a distinguishing feature of the research process. Whenever possible, researchers should plan to collect data electronically, which facilitates the data collection process by eliminating the need to transcribe field notes. Electronic data collection also speeds up data analysis because the researcher need only transfer the data rather than having to manually enter it into the computer. Six tactics are outlined here for using technology in the collection of data.

Tactic 11-1 Using hand-held data input systems

Applications formerly requiring a desktop computer are now available in devices that fit in our hand. Generically referred to as Personal Digital Assistants (PDAs), these devices have significantly expanded the possibilities for field-based researchers. For example, by using software design tools, it is possible for the researcher to create customized data collection applications that essentially turn the PDA into an electronic hand-held clipboard. See Figure 4-2.

Figure 4-2
PalmIII™ connected organizer
Used with permission of
Palm Computing, a 3Com company.

Personal Digital Assistants

PalmPilot™ and Palm III™ organizers
   http://www.palm.com
REX PC Companion
   http://www.franklin.com/rex
Palm-size PC
   http://www.microsoft.com/windowsce/ppc/default.asp
Cassiopeia E-10
   http://www.casio.com
HP 620LX Palmtop PC
   http://www.hp.com/handheld
Newton Message Pad
   http://www.info.apple.com/newton
Psion
   http://www.psioninc.com
Sharp Zaurus, Wizard
   http://www.sharp-usa.com
Velo
   http://www.velo.phillips.com/
CrossPad
   http://www.cross-pcg.com

Vendors selling personal digital assistants and accessories

PDA Page
   http://www.pdapage.com
MobilePlanet
   http://www.mplanet.com
Direct Mobile Com
   http://www.directmobile.com

Tactic 11-2 Using portable keyboards

Portable keyboards were originally developed as a strategy to provide K-12 classrooms with low-cost alternatives to placing a computer on every student's desk. Several devices are available that feature a full-size keyboard with a built-in window (usually about four lines), basic editing tools (delete, copy, paste), and the ability to save multiple files. These devices weigh 2 to 3 pounds and operate for hours on batteries. To transfer information to a computer, simply unplug the keyboard cable from your computer, plug it into the portable keyboard, and press a single key. The files are transferred in seconds and the text will appear in your word processor. At only $200 to $300, these devices seem like an ideal alternative to laptop computers for observational and interview research.

AlphaSmart 2000
http://www.alphasmart.com

A lightweight portable word processor with a built-in LCD screen and full-size keyboard. Enter text as you would on the computer. Store text in one of eight files with up to a total of 64 pages of text. Upload to PC or Macintosh for formatting or printing. See Figure 4-3.

Figure 4-3
AlphaSmart 2000
Used with permission of Intelligent Peripheral Devices, Inc.

DreamWriter 200
http://nts.dreamwriter.com

Features a 16-line screen and built-in disk drive with word processing and printing capabilities, a spell checker, calculator, organizer, and PCMCIA memory card port.

LaserPC5
http://www.perfectsolutions.com/

Includes a word processor, typing tutor, spreadsheet, database, email and calculator. Text may be viewed in one of two formats: 8 lines by 80 characters or 4 lines by 40 characters. Also supports an optional text-to-speech cartridge so writers can listen to what they've written.

Tactic 11-3 Using computer-based data collection tools

Laptop computers are very useful for collecting data. Many different strategies have been developed to use technology to increase the efficiency of collecting data.

Data collection probes

Several manufacturers have created data collection software, sensors, probes, and interfaces that attach to your computer and collect real-time data. See Figure 4-4.

National Instruments
   http://www.natinst.com/labview/
Team Labs
   http://www.teamlabs.com/
Vernier Software
   http://www.teleport.com/~vernier
Texas Instruments, Calculator-Based Laboratory
   http://www.ti.com/calc/docs/cbl.htm

Figure 4-4
A Computer-attachable Probe From Team Labs
Used with permission of Team Labs.

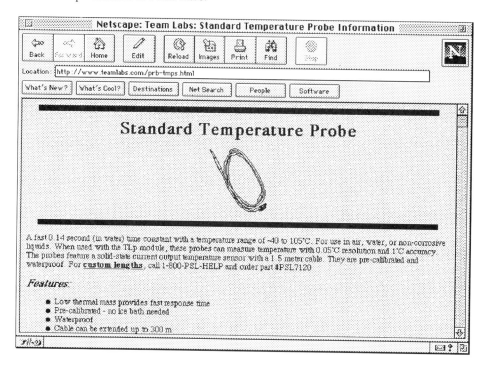

Databases

Databases are valuable tools for creating a place to store and retrieve data. A researcher who has mastered a database program has unparalleled power to collect, store, and review data.

Microsoft Access
http://www.microsoft.com/access/
FileMaker Pro
http://www.filemaker.com
Paradox
http://www.corel.com

Bar coding software

Bar codes are another way to collect and store data. First, the researcher uses software to create the statements, events, or numerical scales that will be used during data collection, and then prints bar code labels. During data collection, the observer uses a bar code reader to scan the appropriate label (usually on a clipboard). Afterwards, the data is uploaded to the computer into a database or spreadsheet.

Optical Data, Lesson Maker
   http://www.opticaldata.com
Bar'n'Coder, BarKoder for Windows
   http://www.ztek.com/Pioneer/pioneer.html
MediaMax
   http://www.videodiscovery.com/vdyweb/school/catalog/
   MAX.htm

Student performance data

Educational researchers are increasingly interested in more holistic appraisal of children's performance rather than simply collecting scores from standardized tests. Several products offer teachers and researchers electronic tools for performance assessment. Portfolio products may be especially useful to researchers engaged in case study research.

The Grady Profile, The Teacher's Portfolio
   http://www.aurbach.com
Learner Profile
   http://www.sunburst.com
Electronic Portfolio
   http://www.scholastic.com
Portfolio Assessment Toolkit
   http://www.ForestTech.com

Tactic 11-4 Using unobtrusive computer-based data collection tools

When research participants interact with the computer, researchers can collect data unobtrusively. This is commonly done through the use of utility software that runs in the background while the user interacts with another program. Unobtrusive data collection software programs capture each keystroke and mouse click entered by the user along with a time code. These tools are ideal for developers in need of transcript monitors and event recording.

WatchIt
    http://www.hi-resolution.com/
Win What Where
    http://www.windowswatcher.com/nidata/p_1692.htm
Historian
    http://www.fortres.com

Another strategy for unobtrusive data collection is to record quicktime movies of all significant events on the screen. When the movies are played back, the observer is able to view the user's interaction with the screen. These tools use a lot of memory but are extremely useful for researchers who design computer interfaces and need to study user behavior and choice-making.

CameraMan
    http:// www.motionworks.com
ScreenCam
    http://www.lotus.com

Tactic 11-5 Using web-based data collection tools

The web increasingly is being used to create surveys that direct research subjects to a specific web page and allows them to interact by selecting responses from a pre-defined list or typing in open-ended responses. This tactic enhances the accuracy of data collection because it eliminates the need to transcribe field notes.

See also	**Tactic 5-4, Locating resources for survey research**

Tactic 11-6 Using voice input

Important developments are being made in the area of voice input: speaking to your computer rather than typing words on a keyboard. Explore the possibilities to determine if the current systems are robust enough for your needs. In most cases, new users speak selected words to "train" the system to recognize their voice. This personalization is an important strategy for improving the accuracy of voice input.

Simply Speaking
   http://www.software.ibm.com/is/voicetype
Dragon Naturally Speaking, Dragon Dictate,
Power Secretary-Macintosh
   http://www.dragonsys.com
Kurzweil VoicePad
   http://www.kurzweil.com/kurzweil/pcapps/voicepad/
   description.html

Wildfire Electronic Assistant
http://www.wildfire.com
> Acts as a personal assistant by allowing you to manage your telephone by speaking into your computer. Initiates calls, automates the process of returning calls, stores 150 contacts, can route incoming calls to other phones, tracks your schedule, and reminds you of follow-up calls or things you need to do.

Speech Machines
http://www.speechmachines.com
> This web site will transcribe your dictation over the telephone or with Voice It's audio files. Early 1998 prices are $29.95 to register, $9.95 per month for four pages of transcribed text, $3.50 for each additional page.

MicNotePad
http://moof.com/nirvana
> A dictation program for the Macintosh. Use the built-in microphone or external microphone, on-screen software buttons for stop, play, record, and skipback. It can be operated in the background of your word processor, making it very convenient to combine transcription, editing, and writing. It makes compressed sound files; four hours of clear audio consumes about 40 mb of harddrive space.

Norcam
http://www.norcam-electronics.com

> Norcam makes a mini-cassette recorder that works in conjunction with Dragon Dictate, transcribing your dictation almost instantly!

Speech Systems Inc.
http://www.speechsys.com

> Speech Systems has developed speech recognition technology that they license to other developers to create new speech recognition applications.

Tactic 11-7 Using digital cameras

Because digital cameras are becoming increasingly sophisticated and less expensive, they hold special promise for researchers. They operate the same as most 35mm cameras, but record the images digitally (still pictures or digital movies) instead of with film. The images are uploaded to your computer where they may be viewed, manipulated, saved, or printed. Common models of digital cameras include:

Casio
> http://www.casio.com

Canon
> http://www.usa.canon.com

Epson
> http://www.epson.com

Kodak
> http://www.kodak.com

Polaroid
> http://www.polaroid.com/products/digital_camera/
>   digcam.html

QuickVideo Transport
http://www.alaris.com

> Plug and play video! Plug QuickVideo Transport into your PC's external parallel port and the other cable into any video source (digital camera, camcorder, VCR) and you are ready to capture motion video or stills on your computer. You can even email your pictures and the receiver will not need any special software to view the images.

Camera Bits
http://www.camerabits.com

> Camera Bits is a software publisher specializing in software for digital cameras. Useful for researchers with unique requirements for their work.

Task #12: Communicating with Field Sites

An outstanding use of technology for the researcher involves improving communication with field sites. Many choices are available depending on the needs of the research team and the level of technology available. Communication may occur through email, web telephone, video conferencing, audio graphics, or collaborative work group software.

Tactic 12-1 Using email

Email is a valuable tool for daily communication with research assistants in the field and centralized staff. As experienced users will recognize, it is a very efficienct tool for resolving problems and a practical means for distributing announcements. Indeed, many principal investigators require all members of the research team to be available via email and establish a regular schedule for exchanging information.

Sophisticated users will also make use of FTP (file transfer protocol) for sending attachments. A field site (e.g., a school) could scan samples of a child's work and forward them as attachments to a remote site for analysis. One common email program that simplifies the process of sending and receiving attachments is Eudora, http://www.eudora.com

---

See also	Tactic 7-2, Using the web as a telephone
	Tactic 7-3, Using video conferencing
	Tactic 7-4, Using audiographics
	Tactic 7-5, Using work group software

CHAPTER 5
Analyzing the Data

For many researchers, the use of technology to enhance research productivity is simply a matter of using computers to analyze data. Unlike some of the other technology tools described in this book, many of the tools in this chapter will be recognized as traditional research tools. As such, they are widely recognized as standards and have proven themselves over a period of years. In this chapter we examine data analysis from both quantitive and qualitative research perspectives using generic and specialized tools.

Task #13 Analyzing Quantitative Data
        Tactic 13-1   Using generic tools for data analysis
        Tactic 13-2   Using specialized data analysis tools

Task #14 Analyzing Qualitatitive Data
        Tactic 14-1 Using generic tools for text analysis
        Tactic 14-2 Using specialized text analysis tools

Task #13: Analyzing Quantitative Data

The use of technology for statistical analysis has long been a tradition in educational research. Many of the tools in this section have generated a considerable industry of accessories: how-to books, critiques of specific analyses, training programs, and so on. As a result, users are likely to find more local support for these products than any other products presented in this book. Two types of statistical analysis tools will be examined: generic and specialized.

Tactic 13-1   Using generic tools for data analysis

Over the past few years spreadsheets and hand-held calculators have increased in power. Spreadsheet programs have formulas for computing mean, media, mode, and standard deviation. Depending on the nature of the data, generic tools may be adequate for researchers who need to report only descriptive statistics.

TI Graphing Calculators
http://www.ti.com/calc/
> Provides descriptions of Texas Instruments's complete line of graphing calculators as well as links to instructional materials and educational resources.

Graphing Calculators
http://www.sharp-usa.com
> Sharp Electronics offers several models of graphing calculators.

Statistical Calculators
http://www.stat.ucla.edu/calculators/
> A very impressive collection of technical programming, this site houses an extensive collection of web-based calculators. Useful for novices and experienced researchers. Visitors with a technical programming background will be interested in exploring the coding and tools used. Maintained by the UCLA Statistics Department. See Figure 5-1.

Figure 5-1
A Web Page Devoted to Statistical Calculators
Used with permissionof UCLA Department of Statistics.

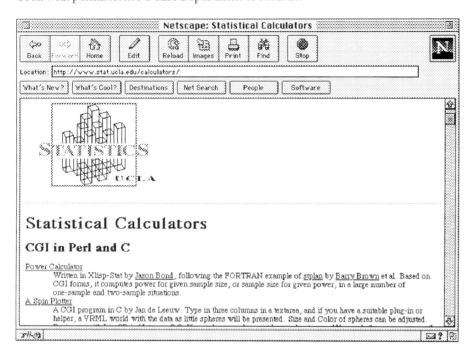

Tactic 13-2   Using specialized data analysis tools

This tactic is traditionally associated with statistical analysis. The technology tools for statistical analysis are well known and come the closest to anything discussed in this book to being considered a standard within the profession. Whereas most products were originally developed for mainframe computers, all now feature microcomputer versions, providing the researcher with considerable power on the desktop.

SAS
http://www.sas.com/

Provides extensive resources supporting the SAS product line as well as BMP software and StatView Software. Use this site to preview software, download demos, view publications, access technical support or training, and check out a variety of links and resources. Of particular value are the search terms that allow you to search the site by Application, Product, or Industry.

Figure 5-2
The SPSS Web Site
Used with permission of SPSS Inc.

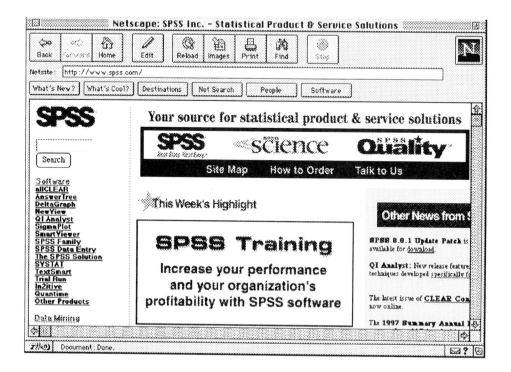

SPSS
http://www.spss.com/

A mega site shown in Figure 5-2. Provides information for all SPSS software products, including allCLEAR, AnswerTree, DeltaGraph, NewView, QI Analyst, SigmaPlot, SmartViewer, SPSS Family, SPSS Data Entry, The SPSS Solution, SYSTAT, TextSmart, Trial Run, In2itive, and Quantime. It also includes downloadable software, resources, links, datasets, whitepapers, and services. Invest some time to explore all that this site has to offer.

Data Desk
http://www.datadesk.com/

A fast, easy-to-use data analysis package, includes powerful tools for exploratory data analysis as well as statistical analysis procedures. Invaluable for probing for a deeper understanding of the patterns, relationships, and exceptions in your data.

MINITAB
http://www.minitab.com

Features descriptions, demos, and resources related to MINITAB statistical software. Download a full-feature 30-day trial version of MINTAB 12 software. This site offers helpful links, product and training information, and technical support.

Mathematica
http://www.wolfram.com

The home of Mathematica, a powerful statistical and visualization tool. Mathematica is a fully integrated environment for technical computing for both development and production in all scientific fields. Its features include advanced numerical computation, algebraic computation, mathematical functions, graphics, programming, notebook interface, and a host of add-on packages for specialized applications. Student versions are also available.

Maple
http://www.maplesoft.com

Maple V 5.0 is a powerful interactive computer algebra and visualization system. Download a free demo. A student version is also available.

Statistical Solutions
http://www.StatSolUSA.Com

Created by the developers of BMP software, this is the home page for Statistical Solutions. Although BMP software is jointly distributed with SPSS, Statistical Solutions currently is the exclusive distributor for Solas for Missing Data Analysis, and nQuery Advisor.

DSTAT Version 1.10
http://www.erlbaum.com/1844.htm

DSTAT is software that was developed for reviewing literature as part of a meta-analysis. Extensive and sophisticated routines for computing study effect sizes (both d and r) and analyzing these values.

Forecast Pro
http://www.palisade.com

Forecast Pro uses artificial intelligence routines to analyze statistical information and predict trends. Users provide the historic data and Forecast Pro analyzes the data, selects the appropriate forecasting technique, builds the model, and calculates the forecast. Palisade creates and markets software for advanced risk and decision analysis. This site has lots of products, information, and resources to explore.

TESTFACT 2.6
http://www.erlbaum.com/2254.htm

> A DOS software program for use in item analysis and test scoring according to the principles of classical test theory. A useful tool for item selection and interpretation of scales, it processes data from binary scored tests, subtests, or scales.

Resampling Stats
http://www.statistics.com

> Describes the methods involved in resampling, drawing repeated samples from the given data, and constructing a simulation. This site features the software program Resampling Stats, along with resources that can be used with this approach to teaching and conducting research. Download a free 30-day trial version.

Task #14: Analyzing Qualitative Data

Qualitative researchers frequently pore over massive amounts of field notes and other text-based data. The technology toolkit that supports the work of qualitative researchers is not as extensive as the toolkit available for quanitative researchers. However, considerable developments have been made in recent years. Two types of tools for conducting qualitative research are examined here: generic and specialized.

Tactic 14-1 Using generic tools for text analysis

Just as the spreadsheet is an essential tool for the quantitative researcher, so is the word processor an essential tool for the qualitative researcher. By entering field notes into the word processor, the researcher can create "tag marks," or codes that can be subsequently searched by using the word processor's "find" command. Also, researchers can search for text that supports a theme they are investigating and copy and paste the information into a second file as a means of extracting evidence supporting their hypotheses. This approach is clearly functional and quite affordable, but researchers may find the need to migrate toward more powerful tools, particularly when they are managing large amounts of information.

Figure 5-3
Home Page of Qualitative Solutions and Research
Used with permission of QSR.

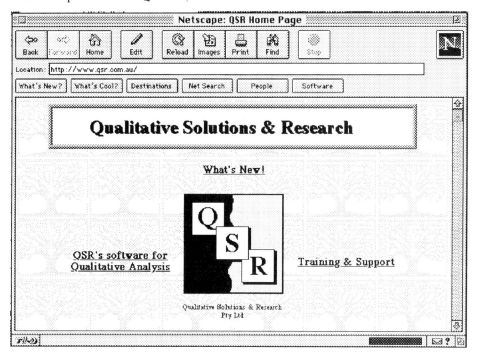

Tactic 14-2 Using specialized text analysis tools

Qualitative researchers have developed specialized tools to assist them in sorting through massive amounts of text-based field notes. The tools in this section are of particular interest to researchers whose data consists of unstructured data from interviews, videotape transcripts, and so on.

QSR NUD•IST
http://ww.qsr.com.au/

Download a demo of this program (Macintosh or Windows) from the QSR Home Page, shown in Figure 5-3. QSR NUD•IST is a popular tool among qualitative researchers, designed for the development, support, and management of qualitative data analysis projects. NUD•IST stands for Nonnumerical Unstructured Data Indexing, Searching and Theorizing (a good thing to know in case you plan on trying to get a purchase order through your institution for nudist software).

The Enthnograph
http://www.qualisresearch.com/

Developed for qualitative social science research, The Ethnograph simplifies the process of categorizing, searching, and sorting data according to a researcher's own analytic scheme. Windows compatible, this product facilitates the management and analysis of text-based data (e.g., diaries, transcripts, focus groups, and field notes). An extremely popular program.

HyperResearch
http://www.researchware.com/

HyperResearch is a product for both Windows and Macintosh that facilitates the analysis of qualitative data. Several unique coding features make this tool easy to learn and easy to use. Visit this site to gain a descriptive and visual overview of how the program works, then download a free demo version to try it out.

Kit–Qualitative Interview and Therapy Analyse
http://www.psy.aau.dk/~dolphin/programs.htm#kit

Kit–Qualitative Interview and Therapy Analyse was developed by Carl Verner Skou, Institute of Psychology, Denmark. This tool was designed to assist qualitative researchers in the analysis of audio data (such as interviews, speeches, and therapy sessions) that have been saved either directly on the computer or transferred from recordings to the computer. This strategy completely eliminates the need for transcription. The data are in the computer allowing you to quickly record your notes and jump from section to section in about one second. An amazing product! Download a free copy.

AQUAD
http://www.aquad.com/

Aquad Five, Software for the Qualitative Analysis of Data, developed at the "Abteilung für Pädagogische Psychologie" of the Institut für Erziehungswissenschaft at the University of Tübingen, Germany. Download a free 30-day demo copy of this software, Windows compatible.

QMethod
http://www.rz.unibw-muenchen.de/~p41bsmk/qmethod/

This page is a mega site for information on the Q-method, a statistical procedure often associated with factor analysis that also has significant application for qualitative researchers interested in assessing subjective topics (i.e., aesthetic judgment). Download free copies of the Macintosh and Windows versions of the QMethod software and access a variety of resources and links on the QMethod. This site was developed and maintained by Peter Schmolck, University of the Federal Armed Forces, Munich.

QDA Software Resources
http://www.ualberta.ca/~jrnorris/qual.html
Judy Norris maintains this excellent site, which is devoted to qualitative research and provides links to qualitative software around the world. Researchers interested in software products in languages other than English will find many useful links to explore.

The Childes Project
http://www.erlbaum.com/893.htm
A textbook, *The Childes Project: Tools for Analyzing Talk,* and a CDROM of analysis tools provide a powerful package of resources for researchers involved in language research (discourse analysis).

ATLAS/ti
http://www.atlasti.de
A comprehensive tool for qualitative analysis of large bodies of text, images, and audio data. Facilitiates the data analysis tasks of selecting, coding, annotating, and comparing. The ease of management and visualization of your data contributes to theory building. It is HTML compatible, enabling you to publish your reports to the web directly from this program. Download a free demo. Windows compatible.

Diction 4.0
http://www.scolari.com/Diction.htm
A Windows product designed to determine the tone of a verbal message using powerful language analysis software that searches a passage for five general features and 35 subfeatures. Provides normative data and produces written reports. Free demo to download.

CHAPTER 6
Reporting the Results

After the data have been analyzed, researchers turn their attention to the final phase of the research process: reporting and disseminating the findings of their research. In this chapter we examine four common tasks: writing, visual representation of data, dissemination of information, and presentation. Eight tactics are described to illustrate the use of technology in conducting these tasks.

Task #15 Consulting Writing Resources
>   Tactic 15-1 Using ready reference tools for writers
>   Tactic 15-2 Using guides to grammar, word use, and style

Task #16 Exploring the Visual Representation of Data
>   Tactic 16-1 Using generic tools for visual representation of data
>   Tactic 16-2 Using creative tools for visual representation of data
>   Tactic 16-3 Using specialized tools for visual representation of data

Task #17 Disseminating Research Findings
>   Tactic 17-1 Disseminating information in print formats
>   Tactic 17-2 Disseminating information via the web
>   Tactic 17-3 Disseminating information in electronic formats

Task #18 Presenting Research Results
>   Tactic 18-1 Using presentation software

Task #15: Consulting Writing Resources

Writing for publication is an acquired skill. It is a myth that some are simply born with the talent to prepare a draft of a manuscript, mail it off to a journal, and have it accepted without revisions on the first attempt. Indeed, most researchers are in need of resources to assist them in the writing process from time to time. In this section we review two types of writing aids: ready reference, and guides to grammar, word use, and style.

Tactic 15-1   Using ready reference tools for writers

With ready access to the web, a writer has an unlimited reference desk at his or her fingertips. This section will assist a writer in locating or verifying specific factual information.

Inkspot
http://www.inkspot.com
A comprehensive site with links to a vast number of writer-related pages on the web.

Writer's Reference Desk
http://www.inkspot.com/ref/refdesk.html
A comprehensive site of ready reference resources. Tons of information only a click away.

Yahoo! Reference
http://www.yahoo.com/Reference
A comprehensive collection of reference materials organized by topics (calendars, dictionaries, encyclopedias, phone numbers, research papers, etc.).

Virtual Reference Desk
http://thorplus.lib.purdue.edu/reference/index.html
A comprehensive collection of reference materials. Developed and maintained by the libraries at Purdue University.

The Writer's BBS
http://www.writersbbs.com
An interactive literature site. Extensive collection of links to other literature sites.

Desktop Reference Agent
http://www.ll.mit.edu/Deskref
This site illustrates research at MIT and the use of agents as a tool for conducting searches. A very interesting resource to explore.

My Reference Desk
http://www.refdesk.com/facts.html
Easy-to-use site. Access links to maps, dictionaries, thesauruses, encyclopedias, government information, and telephone books.

My Virtual Reference Desk
http://www.refdesk.com/outline.html
A wealth of resources organized in a very easy-to-use format.

Xplore Reference
http://www.xplore.com/xplore500/medium/reference.html
A directory-based list of ready reference web resources.

The Internet Public Library Reference Center
http://www.ipl.org/ref
This site allows users to consult reference materials using categories commonly used in libraries for organizing disciplines. A useful interface for those familar with their topic.

StudyWEB
http://www.studyweb.com
Designed for students but filled with research-quality links.

Almanac
http://www.almanac.com
The electronic equivalent of the *Old Farmer's Almanac*.

Atlapedia
http://www.atlapedia.com/index.html
Key information on every country in the world.

The World Factbook
http://www.odci.gov/cia/publications/factbook/index.html
A publication of the Central Intelligence Agency, provides complete information about all countries of the world.

John December's List of Essential Resources
http://www.december.com/cmc/info/

A popular resource for Internet trainers and students to use to learn about the Internet. Developed and maintained by John December, it is a comprehensive collection of information sources about the Internet and computer-mediated communication.

Finding Data on the Internet
http://nilesonline.com/data/

Robert Niles provides strategies and links to locating statistical data online. Designed for journalists seeking to cite specific facts.

Ask Jeeves
http://askjeeves.com

This site encourages users to ask questions in natural language (sentence format). Using principles of artificial intelligence, the search engine attempts to make matches between your inquiry and reference resources. A very interesting site to explore.

Information Please
http://www.infoplease.com/

A collection of well-known almanacs, encyclopedias, and dictionaries, including Information Please. Users may browse or search.

Thesaurus of Geographic Name
http://www.ahip.getty.edu/vocabulary/tgn.html

A database of almost 1 million place names and 900,000 places around the world. Entries feature latitude and longitude, place names and types, and sources of information. Developed by the Getty Information Institute.

SYMBOLS.com–Encyclopedia of Western Signs and Ideograms
http://www.SYMBOLS.com/

This site is an electronic version of *Thought Signs,* a book by Carl G. Liungman, an encyclopedia of graphic symbols. More than 2,500 Western signs arranged in 54 groups based on the graphic characteristics. Searches may be conducted on the graphic symbol or meaning.

Copyright Registration for Freelance Writers
http://www.asja.org/cwpage.htm

Provides information for copyrighting your articles.

Figure 6-1
Determining the Appropriate Bibliographic Format for
Citing Electronic Information
Used with permission of Nancy Crane.

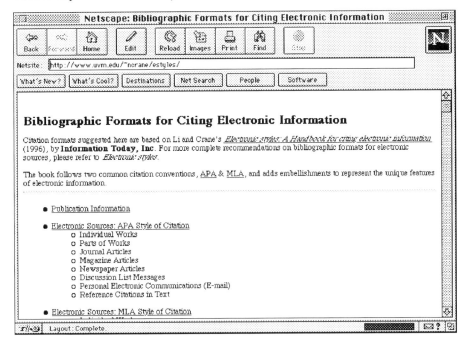

Tactic 15-2 Using guides to grammar, word use, and style

A number of style guides are available online to assist writers with questions of word usage, grammar, and style.

Guides to style

Bibliographic Formats for Citing Electronic Information
http://www.uvm.edu/~ncrane/estyles

This site covers both APA and MLA style guide enhancements for citing electronic information. Based on Li and Crane's *Electronic Styles: A Handbook for Citing Electronic Information* (1996, Information Today, Inc.). See Figure 6-1.

MLA Style: Citing Sources from the World Wide Web
http://www.mla.org/main_stl.htm

This site is sponsored by the Modern Language Association (MLA) and contains the the MLA authorized guidelines for citing internet resources.

Web Extension to the American Psychological Association Style (WEAPAS)
http://www.beadsland.com/weapas/
Proposed standards for referencing online documents in scientific publications.

Beyond the MLA Handbook: Documenting Electronic Sources on the Internet
http://falcon.eku.edu/honors/beyond-mla
By Andrew Harnack and Gene Kleppinger of Eastern Kentucky University, a thoughtful essay about the problems and practical issues of documenting electronic sources. Includes many hypertext links to resources and a style guide.

Style Guide for Online Hypertext
http://www.w3.org/hypertext/WWW/Provider/Style/
This style guide offers guidance on preparing information for web publication. An essential reference for webmasters.

Strunk and White's Elements of Style
http://www.columbia.edu/acis/bartleby/strunk/strunk.html
The copyright on this classic text has expired. Consult it on the web.

The CopyEditing Style FAQ
http://www.rt66.com/~telp/sfindex.htm
The informed opinions of members of the copyediting-l mailing list were organized into a question-and-answer format about style and usage. A very accessible format for practical advice.

Guides to word usage

A Web of Online Dictionaries
http://www.bucknell.edu/~rbeard/diction.html
A mega-site containing links to dictionaries in more than 130 languages. Developed and maintained by Robert Beard of Bucknell University.

Random House Unabridged Dictionary
http://www.answers.com/reference.cgi
Consult the electronic version of this popular dictionary.

Dictionary.com
http://www2.dictionary.com

This home page uses the DICT Server version of *Webster's Revised Unabridged Dictionary* (G & C. Merriam Co., 1913, edited by Noah Porter).

WordNet
http://www.cogsci.princeton.edu/~wn/

WordNet is a lexical database, a powerful lexical reference system combining components of dictionaries, thesauruses, and theories of psycholinguistic lexical memory. Words are defined and grouped into sets of synonyms. Developed by the Cognitive Science Laboratory at Princeton University, the site provides information on how to download copies of WordNet.

Research-It!
http://www.itools.com/research-it

An excellent collection of language tools. Check out the universal translator that lets you key in a word to find its equivalent word in another language.

Colorado State University Online Writing
Center Reference Materials
http://www.colostate.edu/Depts/WritingCenter/reference.htm

Practical information on seven topics related to critical writing. Developed and maintained by the Center for Research on Writing and Communcation Technologies at Colorado State.

Guides to grammar

Online Writing Labs (OWL)

Online writing labs provide a variety of reference materials as well as how-to guides for writers. Many OWLs are maintained by the writing department at a university as a means of assisting students in developing their writing skills. Each site has resources on grammar, but be sure to explore the wealth of information on other subjects. Also note that most OWLs will accept questions and will respond via email with their advice.

Purdue University OWL
http://owl.english.purdue.edu
The University of Maine's Online Writing Center
http://www.ume.maine.edu/~wcenter/resource.html
The Writing Center at the University of Wisconsin-Madison
http://www.wisc.edu/writing/index.html
University of Michigan Online Writing Lab
http://www.lsa.umich.edu/ecb/OWL/Resources.html

Online English Grammar
http://www.edunet.com/english/grammar/index.html
> A detailed technical site concerning English grammar. Be sure to check the Digital Education Network's English Language Practice Pages.

Task #16: Exploring the Visual Representation of Data

Increasingly, researchers are concerned with finding new ways to communicate the meaning of their data. The field of visual literacy is helping us understand how to visually present data in powerful ways. For this task, three tactics are presented that provide generic and specific tools for visual representation of data and creative tools that allow freehand design.

Tactic 16-1 Using generic tools for visual representation of data

All spreadsheet programs have a graphing component to allow the user to examine the data visually in a chart. A pie chart can be changed into a bar chart with a click of a button, but it is important to understand why one chart is useful for illustrating a budget and another chart is useful for achievement scores over time. The tools in this section will be useful for many applications.

DeltaGraph 4.0
http://www.spss.com/software/science/deltagraph/
> Provides more than 50 mathematical, statistical, and data analysis functions along with 70 chart types and 200 chart styles.

SPSS Diamond 1.1
http://www.spss.com
> A tool for exploring data and discovering patterns. Imports files from a variety of common statistical file formats, such as ascii, SPSS, Systat, Lotus, and dBase.

KaleidaGraph
http://www.synergy.com
> KaleidaGraph is a data analysis and graphing program for Windows and Macintosh computers. Handles large data sets well and supports robust curve fitting.

Analytica 1.0.1
http://www.decisioneering.com

A modeling tool that allows users to build a visual layout of the variables, their relationships, and the methods for computing the results. Provides significantly more power than a spreadsheet.

Visio Professional 4.5
http://www.visio.com

A powerful tool for creating diagrams. Has a feature to export diagrams as HTML pages.

Inspiration
http://www.inspiration.com

Inspiration is a software product for developing concept maps. Very useful for brainstorming.

Timeliner
http://www.teachtsp.com/classroom/timelinerOnline/
tlineronline.html

A software program for creating and labeling historical data in a time line format. Time lines can be published to the web. Download a free demo.

Tactic 16-2 Using creative tools for visual representation of data

Graphic artists have found a variety of methods to make data meaningful. Indeed, the hugely popular *USA Today* data charts have done much to advance the proposition that data need not simply be charted to be effective.

Paint programs

The most common tools for creatively illustrating data are paint programs. The user is given a palette of tools (line, shapes, colors, effects) with which to draw and paint freehand images.

Adobe Illustrator
http://www.adobe.com/prodindex/illustrator/prodinfo.html

The standard design and illustration tool. Allows users to create visual images with complete control over all aspects of the image, including layers and special effects. An essential tool for web design and print materials.

CorelXARA
http://www.corel.com/products/corelxara2/index.htm

A powerful graphics tool designed to create graphics for web publishing. CorelXARA 2.0 is available only as a downloadable product.

Painter 5
http://www.metacreations.com/products/

Software for Macintosh and Windows by MetaCreations Corp. Download a free demo copy. Features an extensive set of tools and brushes for designing and editing images.

Clip art

For the artistically impaired, clip art and stock photos are invaluable. These professionally designed images are used simply by copying and pasting.

Art Today
http://www.arttoday.com/

Browse this site and see for yourself the value of an online image library. Subscribe for one annual fee and you'll have access to 600,000 clip-art images, 50,000 photographs, 5,000 web-graphics, bullets, horizontal rules, icons, and 1,700 TrueType fonts. There is no licensing fee, which means that users are free to reproduce the images in their work without additional charges.

MasterClips Premium Image Collection
http://www.imsisoft.com

A CDROM for Windows featuring 303,000 images from several clip-art collections.

Inki's Clipart
http://www.inki.com/clipart

A collection of public domain clip art. Free to download and use on your personal web site.

Cool Graphics on the Web
http://little.fishnet.net/~gini/cool

A mega directory of free graphics on the web. This page takes a long time to load, but it's worth the wait. The directory provides a use organizer for labeling a variety of image styles.

The Clip Art Directory
http://www.clipart.com

Another mega collection of clip art, this site serves as a directory of links to other web sites. The organization of sites by names is not always helpful, but browsing the extensive collection is a great start.

The Iconfactory
http://www.iconfactory.com/

Freeware icons to dress-up your desktop. The resources at the Icon Factory are very useful if you are creating software or web sites.

Photo Impact 4
http://www.ulead.com

Visit this site by Ulead Systems and explore their image software. Of particular interest is the product Ulead Impact 4, an inexpensive image editor. Free downloads of selected software products available.

PhotoCD
http://www.kodak.com/daiHome/products/photoCD.html

Visit this site to explore how to create a PhotoCD from images you capture using your 35mm camera. Describes the production process as well as the many uses of your digital images once the PhotoCD is produced. The PhotoCD is a viable alternative to digital cameras in some applications.

Riptide Photo Studio
http://www.vorton.com

This graphics utility is designed to view, convert, and edit graphics from more than 40 different file formats. Conversion software like this can be invaluable as you work with images in multiple formats.

Tactic 16-3 Using specialized tools for visual representation of data

Occasionally, a researcher will find a need to use specialized tools to visualize data. Specialized tools may include modeling programs, geographic information systems, and mapping programs.

Modeling

Crystall Ball Pro
http://www.decisioneering.com

An interactive program that helps identify risks in decision making. Users enter data, define and creat graphical representations of uncertain values known as "assumptions," and generate "what-if" scenarios. The program generates forecasts based on the results. Available from Decisioneering, Inc., Denver, Colorado, for both Windows and Macintosh.

Visable Analyst
http://www.visible.com

Often bundled with engineering and design textbooks, Visable Analyst is a software product that is used in the design, development, and documentation of new products. Download a free demo copy.

HiQ
http://www.natinst.com/hiq/

HiQ is an interactive environment for problem solving. Users analyze, visualize, and document data using a notebook methodology that integrates interactive analysis, data visualization, and report generation. Produces a container for a complete technical document that can be distributed.

Precision Mapping Streets 3.0
http://www.chicagomap.com/

A CDROM for Windows allows users to search and display views of every street segment, waterway, railway, and political boundary in the United States. Users can search by country, city, zip code, street, landmark, or area code and create custom maps.

Proximity 3.0
http://www.decisionmark.com

A software product that comes with detailed U.S. maps, and demographic and economic data that enables users to plot their data on maps to create graphical representations. A valuable tool for identifying underserved areas and cluster areas.

ArcView
http://www.esri.com
> A geographic information system for schools and libraries that enables the average citizen to explore the power of interactive mapping and analysis. Useful for examining geographic, social, economic, political, and environmental information for answering practical questions. A complete desktop system for storing, modifying, querying, analyzing, and displaying information about places as near as your neighborhood and as far away as the south pole.

Task #17: Disseminating Research Findings

As researchers prepare a report, they give considerable thought to their audiences. Manuscripts are primarily written and submitted to professional journals; they are also prepared for presentation at conferences. Increasingly, researchers are being asked to translate their findings into formats that can readily be used to improve practice. Three tactics have been identified to assist researchers in fulfilling their responsibilities for information dissemination: disseminating information in print, via the web, and in electronic formats.

Tactic 17-1 Disseminating information in print formats

The word processor is a valuable tool for most researchers, allowing them not only to input text easily, but also to copy and paste between documents to create such products as research summaries. The word processor also serves as an archive to store and print copies of previously written reports (see Tactic 8-1).

However, when the visual design of print information becomes complex, the limits of word processing may become apparent. In such cases, a researcher will find a need for desktop publishing software. Several popular desktop publishing programs for creating newsletters, flyers, and research summaries include:

QuarkXPress
> **http://www.quark.com**

PageMaker
> **http://www.adobe.com/prodindex/pagemaker/main.html**

Microsoft Publisher
> **http://www.microsoft.com/products/**

Plug-ins for Mathematical Symbols
http://www.w3.org/Math

MathML is a Mathematical Markup Language that works as a web browser plug-in to enable web pages to natively display mathematical symbols.

Blue Sky Research
http://www.bluesky.com

An interactive desktop publishing system for producing scientific journals and textbooks. Uses the Tex typesetting language.

Globalink
http://www.globalink.com

Globalink has a series of products that facilitate language translation in Spanish, French, Portuguese, German, Italian, and English. Power Translator software translates from English to one of four langugages; Web Translator translates web sites written in Spanish, German, or French to English; and Language Assistant software provides assistance as you engage in language translation activities.

The Design and Publishing Center
http://www.graphic-design.com

New to desktop publishing? Information about graphics, typography, publishing, and printing. Useful for beginners and intermediate users alike.

The Xerox Document Center
http://www.documentcentre.com

Explore this site to review Xerox's vision for the possibilities associated with seamless integration of information across the desktop: copy, print, scan, or fax. Describes strategies and products for connecting your computer to a publishing network.

Figure 6-2
Example of a PowerPoint Presentation Published on the Web
Used with permission of Cheryl Wissick.

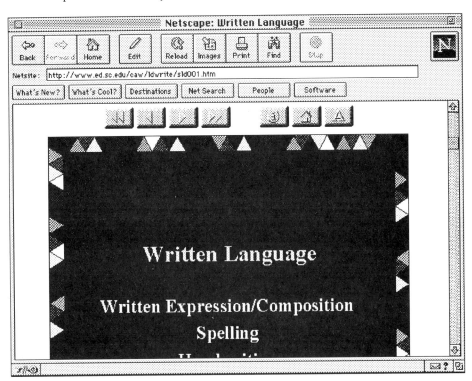

Tactic 17-2 Disseminating information via the web

Using the web to disseminate information about a research project is becoming more popular because of its ease, flexibility, and cost-effectiveness. Visitors to your page will peruse the information you have available. Documents may be viewed, printed, or downloaded.

Written Language
http://www.ed.sc.edu/caw/ldwrite/index.htm

Did you know that Microsoft PowerPoint presentations could be published on the web? Visit this site to experience a web-based presentation (see Figure 6-2). This technique has important implications for the dissemination of research conference presentations. Developed by Professor Cheryl Wissick, University of South Carolina.

Adobe Acrobat
http://www.adobe.com/acrobat/readstep.html

Have you ever brought a document that you viewed on someone else's computer back to your office, only to find that you couldn't view it because you didn't have the program used to create the file? This problem is solved by Adobe Acrobat, which allows you to use your existing electronic files and turn them into pdf files (portable document files). This process maintains the format and style of your document but enables anyone to open and view the file by using the free program, Adobe Acrobat Reader.

Design and development

Style Guide for Online Hypertext
http://www.w3.org/hypertext/WWW/Provider/Style/

This style guide offers guidance on preparing information for web publication. An essential reference for webmasters.

Web Pages That Suck
http://www.webpagesthatsuck.com

Learn about web design by critiquing web pages in need of a make-over.

Claris HomePage 3.0
http://www.clarishomepage.com/products/claris/clarispage/ chpbeta_hdr. html

Download a free evaluation copy of this commerical product. Claris HomePage is a tool for creating home pages on the web.

American School Directory, Web Site Advisor's Guide
http://www.asd2.com/resource/advisor.html

A comprehensive guide to planning and preparing a school web site. Provides a useful overview for any new webmaster.

Developer Shed
http://www.devshed.com

Source for web news, tutorials, and design tools for users at all levels.

Webdeveloper.com
http://www.webdeveloper.com/

A site sponsored by Mecklermedia to provide the latest news in web developments. Features a library, Q & As, and a developer's forum.

Figure 6-3
A Field Guide to Institutional Research Home Pages
Used with permission of John Milam.

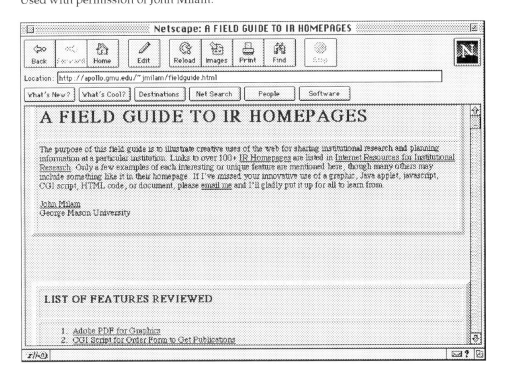

A Field Guide to Institutional Research Home Pages
http://apollo.gmu.edu/~jmilam/fieldguide.html

This site profiles web-design features and offers links to over 100 institutional research home pages to illustrate the design principle in practice. An inspirational and instructive resource for web designers of all skill levels. See Figure 6-3.

HTML Goodies
http://www.htmlgoodies.com/

A huge collection of HTML-related tutorials for beginning and advanced web developers. Includes a library of more than 450 Javascripts. Developed and maintained by Joe Burns.

devSearch
http://www.devsearch.com/

This site offers a meta search engine for simultaneous searching of 23 sites related to web development.

ImageVice 1.1
http://www.boxtopsoftware.com/

ImageVice is a color reduction technology that compresses graphic file size (up to 70 percent) to deal with the problem of lengthy download times. Available for Macintosh and Windows. Download a fully functional demo of ImageVice 1.1, a commerical product.

National Association of Webmasters
http://www.naw.org

A professional organization dedicated to support the work of webmasters. Features an extensive collection of resources for new and experienced webmasters. It also sponsors a mentoring program for new webmasters.

Web demographics

Much is written about the viability of the web as a vehicle for information dissemination. To assess its current efficacy as a tool for your research, you may need to conduct an analysis of the current demographics associated with web users.

Several sites monitor the growth of the Internet, report on the ever-changing demographic profile of users, and analyze trends:

Matrix Information and Directory Services, Inc.
http://www.mids.org
Georgia Tech University's WWW User Surveys
http://www.gvu.gatech.edu/user_surveys/
Internet Demographics
http://calafia.com/classroom/demographics.htm
Survey-Net
http://www.survey.net/
Hermes Project
http://www.umich.edu/~sgupta/hermes/

Hosting a web site

Well Engaged
http://www.wellengaged.com

A leading supplier of software and services for building web communities. Entry-level sites support web-based events and discussions for between 100 and 500 active users, and costs $250 a month with a one-time setup fee $3,500. The software can also support more than 50,000 active users.

Lotus LearningSpace 2.0
http://www.lotus.com

LearningSpace is a comprehensive course-creating and management tool. The system utilizes a Domino web server to provide distance learning applications. Includes five modules: Schedule, MediaCenter, CourseRoom, Profiles, and Assessment Manager.

FirstClass Intranet Server
http://www.softarc.com/

SoftArc Inc. offers a special verion of its FirstClass Intranet Server for creating work group environments. FirstClass software is a standard in educational networking that provides excellent cross platform compatibility for email and intranets.

Netopia Virtual Office
http://www.netopia.com/software/nvo/

Designed for small offices, this program combines tools for Internet collaboration and remote control capabilities in a virtual office interface. Establish a pre-built web site with your own URL as part of geocities.

Site maintenance

Engage
http://www.engagetech.com

A suite of products (Engage.Journal, Engage.Portrait, Discover, and Engage.Knowledge) to assist in collecting and analyzing web-site visitor data while serving customized content.

FireFly
http://www.firefly.net

FireFly licenses technology that enables web sites to customize content based on user preferences. Profiles are kept as "virtual passports" that users carry from site to site. An innovative strategy to address concerns about privacy associated with web commerce while simultaneously helping businesses forge highly personalized services for their customers.

SiteSweeper 1.0
http://www.sitetech.com

Runs a full diagnostic checkup on a web site on more than 80 separate defects in seven major quality areas. Utilizes a "crawler" to gather data on all text, image, audio, and video files, including size, time they were last modified, and total download size of each page. An essential tool for webmasters, it is a Windows software product.

Bongo
http://www.marimba.com

As network managers know, updating software and data files across an organization can be a tedious task. Bongo is a software tool that utilizes the Castanet system for providing users with access to the most current version of software and datafiles regardless of where they exist on the network.

Net-It Now
http://www.net-it.com

This product is designed to facilitate web publication of print documents. An alternative to the popular pdf format used by Acrobat.

Web Position Analyzer
http://www.webposition.com/iworld

Software to monitor your site's search engine rankings in order to help you improve your site's visibility.

InfoPress
http://www.castelle.com/

InfoPress is a family of software products from Castelle Inc. that brings together requesting and delivering information from a central library using the most common delivery systems: telephone, fax, email, and web. Using a message stored in a single format, the software allows users to select how they would like to receive information from your site: voice, fax, email, or the web. Subsequently, the software converts the message via the voice module for fax on demand, email via email, or fax, or converts to HTML for web delivery.

Tactic 17-3 Disseminating information in electronic formats

Researchers have a variety of options when considering the distribution of information in electronic format. Disks are an excellent option when the information will fit on a few diskettes. However, the cost associated with producing your own CD has declined dramatically in the last few years. As a result, mastering and distributing CDs may be extremely cost-effective. Finally, developments in the area of desktop video also make distributing information electronically a viable option.

Multimedia portfolios

Educational researchers are likely to gather a variety of data from their subjects. Multimedia portfolios are useful for creating electronic case studies. In addition to providing an efficient means for archiving data, these products are invaluable in presenting the outcomes of your research.

Digital Chisel
http://www.pierian.com
The Grady Profile
http://www.aurbach.com
HyperStudio
http://www.hyperstudio.com

Folio 4
http://www.folio.com

A series of products that allow users to create their own information bases using Folio Views software. A license allows the developer to distribute runtime versions of the custom databases without additional royalties.

3D Fax
http://www.infoimaging.com

Allows Windows users to compress and encode any computer file into a printable "infoimage" that can be faxed to other users. Recipients then scan and decode images using the 3D Fax freeware. The company claims its programs can compress up to 50 pages of text onto a single-page fax, giving users obvious savings in fax transmission time. Also described are innovative applications for transmitting security-sensitive documents, electronic files (e.g., voice, music, computer programs), and ideographically-based languages (e.g., Japanese).

Figure 6-4
Home Page of Storyspace™
Used with permission of Eastgate Systems, Inc.

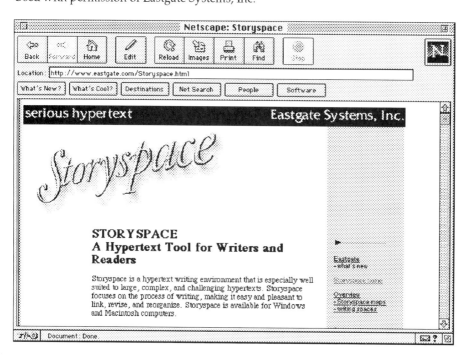

Storyspace™
http://www.eastgate.com/Storyspace.html

Visit this site and learn about Storyspace, a hypertext writing environment, shown in Figure 6-4. Available for both Windows and Macintosh formats, this product enables authors to create large complex stories with hyptertext links. Documents that are produced with Storyspace may be distributed royalty-free as stand-alone programs or published to the web. This tool has considerable potential for the research community as they explore media options for disseminating their work. Developed by Eastgate Systems, Inc.

The CD Information Center
http://www.cd-info.com

Interested in publishing your own CD? This site has information about the CD industry and many applications of CD technology.

The Multimedia and CDROM Directory
http://www.macmillan-reference.co.uk/

Combines the two print publications *Marketplace* and *Titles* to create a guide of nearly 14,000 companies around the world involved in the production of CDROMs. Available on CDROM for Macintosh and Windows, this directory monitors more than 24,000 products.

Audio Images Interactive Group
http://www.bway.net/aiig

Describes a new technology: CD-enhanced. Described as enabling a whole new level of interactivity on CDs than has been traditionally possible. The CD-enhanced format lets you play video, audio, graphics, text, biographies, interviews, documentaries, live performances, and more on the same CD.

CD-Rewriteable Drive
http://www.philips.com

The Phillips CD-Rewriteable Drive allows user to create their own CD. This technology has become increasingly affordable in recent years, meaning that most medium and large projects can now consider creating their own CDs.

Signature
http://www.fargo.com

A special printer created for printing graphics and text on CDs you develop.

Snappy
http://www.play.com

Snappy is a versatile tool for video editing. Attach this device to your computer (Windows compatible) and edit your video.

Task #18: Presenting Research Results

Presenting the results of one's work at a professional conference is a common task for researchers. A variety of products are available for assisting in this task.

Tactic 18-1 Using presentation software

Presentation software products allow the user to create high-quality visual presentations by designing a series of screens that are presented in the form of a slide show. The presenter controls the rate of presentation with the mouse as the screen image is projected from the computer onto a large screen through the use of an LCD panel or projection system. Presentations may be distributed on disk for others to "play" on their personal computer, or published to the web (see page 131). Common presentation software products include:

PowerPoint
 http://www.microsoft. com
Astound
 http://www.goldisk.com
Persuasion
 http://www.adobe.com
Presentations
 http://www.corel.com

Presenters University
http://www.presentersuniversity.com
 Interested in improving your presentation skills? Visit this free site sponsored by Presentation University. Allows a presenter to develop skills in delivery, content, and visuals, Utilizes the Power of Wow! model.

1998 AERA Conference Papers
http://aera.net/meeting/papers98.html
 Since 1997, AERA has assisted presenters in offering electronic copies of conference papers. This page offers links to papers from the 1998 conference. Watch for future postings preceding each annual conference at http://aera.net/meeting/

Research Presentations on the Internet
http://www.ualberta.ca/~jrnnorris/research.html
 This page by Judy Norris, University of Alberta, offers links to virtual poster sessions, electronic theses and dissertations, and other scholarly electronic presentations of research.

CHAPTER 7
Research Administrator

Students often begin their research careers by collecting data for senior researchers. Student researchers assume additional responsibilities as they design and conduct their investigations with faculty supervision. Career researchers may begin their work as a member of a large research team and eventually move into the role of principal investigator, or perhaps director. In each case, whether the research project is large or small, the researcher is faced with any number of management duties. In this chapter we examine 10 tasks involved in the daily work of a research administrator and consider 21 tactics for using technology to enhance productivity:

Task #19 Managing Information
　　　Tactic 19-1 Creating electronic text
　　　Tactic 19-2 Creating databases
Task #20 Managing Projects
　　　Tactic 20-1 Using project management software
　　　Tactic 20-2 Using time management software
　　　Tactic 20-3 Meeting technical requirements
Task #21 Maintaining Financial Records
　　　Tactic 21-1 Using financial recordkeeping software
Task #22 Selecting Mailing Systems
　　　Tactic 22-1 Mailing and tracking systems
Task #23 Purchasing Supplies
　　　Tactic 23-1 Purchasing office supplies
Task #24 Purchasing Equipment
　　　Tactic 24-1 Purchasing new equipment
　　　Tactic 24-2 Purchasing upgrades
　　　Tactic 24-3 Purchasing used equipment
Task #25 Obtaining Technical Support
　　　Tactic 25-1 Obtaining phone support
　　　Tactic 25-2 Using troubleshooting tools
　　　Tactic 25-3 Accessing web sites that offer technical support
Task #26 Facilitating Travel Planning
　　　Tactic 26-1 Making reservations
　　　Tactic 26-2 Exploring a city
　　　Tactic 26-3 Checking on the weather
　　　Tactic 26-4 Working while you travel
Task #27 Preparing Disaster Recovery Plans
　　　Tactic 27-1 Using backup systems
　　　Tactic 27-2 Planning for disaster recovery
Task #28 Participating in Technology Transfer Processes
　　　Tactic 28-1 Facilitating technology transfer

Task #19: Managing Information

One result of the Information Age is the exponential growth of information. This uncontrolled growth is particularly problematic for researchers and other knowledge workers because the basic tools for navigating this information glut are fairly primitive. Compounding the problem, information is created, manipulated, and stored in a variety of formats. To manage information, research administrators will often acquire tools that facilitate the creation, manipulation, and storage of electronic text.

Tactic 19-1 Creating electronic text

Creating and storing information in an electronic format is important to the researcher because of the ease in which the text can be manipulated–it can be incorporated into a manuscript, printed as a research protocol, emailed to other team members, posted on an intranet, and so on. Several tools and strategies will aid in creating, storing, and manipulating electronic text.

Essential for the creation of electronic text is to ensure that all research project staff members have access to a word processor, the core tool for generating text. See also tactics 11-1 and 11-2 for products that facilitate text generation in the field.

PaperPort
http://www.visioneer.com

Information on paper can be transfomed to electronic text by any one of a number of new desktop scanners that allow you to simply insert a piece of paper through a roller and have the computer scan the text or graphic into your favorite program. The PaperPort is a product distinguished by ease-of-use and low cost. Available for both Windows and Macintosh computers.

Email Attachment Opener, MacOpener
http://www.dataviz.com

Utility software for Windows 95 and Macintosh respectively that will open any email file attachment. By pasting text into an email message or sending it as an attachment, the research team reduces its reliance on paper. Email has many other advantages: users can use email to create documents collaboratively, distribute information quickly, print a paper copy, integrate the information into a personal filing system, and re-use text without retyping.

Adobe Acrobat
http://www.adobe.com/prodindex/acrobat/readstep.html

A utility program that uses your document to create a pdf (portable document format) file. This standard file format can be read by anyone with an Adobe Acrobat Reader (free) and allows information developers to create information using the software of their choice and capture it in a format that readers can view or print but not change. Essential for the dissemination activities of any research project.

Can Opener
http://www.abbottsys.com

Utility software that enables the Macintosh user to open and view a document file without requiring a copy of the software that was used to originally create the document.

File Compression Utility Programs
http://www.pkware.com
http://www.aladdinsys.com
http://www/winzip.com

Sometimes a research team will develop information or a software product and want to make it available to others electronically, but the file's size makes it prohibitive to download. Compression software was made to solve this problem. Self Extractive Applications (SEA files for the Macintosh) and Zipped (ZIP files for Windows) are created by compression utilities software to speed the transmission time in downloading (or to fit it onto a disk). The user simply clicks on the SEA or ZIP file and the file initiates an automatic decompression and installation procedure. PKZip for Windows is the standard for Windows machines and StuffIt is the standard on the Macintosh. For those compressing the file, this is an intermediate-level task, but compressed files are very easy to download and use, even for the novice.

Net-It Central, Net-It Now!
http://www.net-it.com

Another information management strategy involves creating an intranet in which your research team can store and share information via the web. Net-It Central turns your intranet into a document-sharing center. New-It Now! is a web-publishing system that converts files created with standard office software into documents that are ready for publishing on the web.

Tactic 19-2 Creating databases

Databases and electronic forms enable the research administrator to collect and store information systematically and produce reports in a timely fashion. Databases are simply software programs that allow users to create forms for entering information and subsequently search and sort the information. Mastery of a database program can take some time but having information available at your fingertips is an invaluable resource for the research administrator.

Microsoft Access
   http://www.microsoft.com/access/
FileMaker Pro
   http://www.filemaker.com
Paradox
   http://www.corel.com

Another way to ensure that information is available electronically is to transfer your paper forms to an electronic format. Several products are available to assist you in designing electronic forms. Again, once the data is entered, the information can be printed, faxed, or emailed to others.

JetForm
   http://www.jetform.com/
FormTool
   http://www.imsisoft.com/formtool/index.html

Task #20: Managing Projects

The grant writer must outline a plan for managing the resources that will be provided by a grant and align activities with personnel and time lines. This provides the reviewers and the funding agency with some level of assurance that considerable thought has been devoted to ensuring that the project will be successful. Three tactics are described here: software for project management, time management software, and meeting technical requirements.

Tactic 20-1 Using project management software

Project planning software can be used to design management tools such as GANT charts and PERT charts, for monitoring projects, people, resources, and time lines. These types of tools are essential for successful grantwriters who must manage several research projects simultaneously. Of course, they are invaluable in the proposal development process as well.

Microsoft Project
    http://www.microsoft.com/project/
Fast Track Schedule
    http://www.aecsoft.com/
IntraPlan
    http://www.intraplan.com/
Visual Staff Scheduler Pro 3.0
    http://www.abs-usa.com

Tactic 20-2 Using time management software

At the center of most personal productivity training is learning how to manage a daily calendar. A variety of products have been designed to assist users in managing their daily calendar, appointments, and to-do lists. In many cases, you will find an electronic equivalent of your favorite productivity system. Sometimes these products are referred to as Personal Information Managers (PIMs).

DayRunner
    http://www.dayrunner.com
Lotus Organizer
    http://www.lotus.com/home.nsf/welcome/organizer
Claris Organizer
    http://www.claris.com/products/claris/clarisorganizer/
    clarisorganizer.html
InfoSelect 3.0
    http://www.miclog.com

Remind U-Mail
http://calendar.stwing.upenn.edu/
    Here's an interesting web application. Register as a user of this site and you can create a calendar of reminders. Simply enter the day of an event and a brief description of it, and the service will send you an email reminder on that day.

| *See also* | Tactic 11-1, Using hand-held data input systems |

Tactic 20-3 Meeting technical requirements

One critical area of concern for the research administrator involves ensuring that the project is in compliance with the technical requirements of the funding agency. Although this is one area in which experience is invaluable, the new research administrator can turn to several resources for assistance.

Education Departments Grants Administrative Regulations (EDGAR)
http://gcs.ed.gov/grntinfo/edgar.htm

> Regulations concerning the administrative requirements for managing projects funded by the U.S. Department of Education.

NIH Guide to Grants and Contracts
http://www.nih.gov/grants/guide/index.html

> Sponsored by the National Institutes of Health, this site provides a wealth of information about grants, contracts, and other items of interest to researchers.

Next ERA
http://www.rams-fie.com/software/index.htm

> Next ERA is an electronic grants management system. Offers modules for the efficient and effective management of a grant spanning the pre-award and post-award process.

FedLaw
http://www.legal.gsa.gov

> A site run by the General Service Administration to improve access to a variety of sources on federal law. Features a variety of indexes and search engines.

NonProfit Gateway
http://www.nonprofit.gov

> Designed to give nonprofit agencies one-stop access to federal agencies.

C-SPAN
http://www.c-span.org

> The C-SPAN web site captures all developments on C-SPAN and C-SPAN2. You'll find a daily schedule of the two channels as well as roll call votes, and other information. Perfect for monitoring budget bills, committee debates, and so on.

Nolo's Legal Encyclopedia
http://www.nolo.com/briefs.html

The web site of Nolo Press Self-help Law Center, publishers of software and books. The Encyclopedia is an invaluable compendium of legal issues related to small businesses, patents, copyright, trademarks, will and estate planning, and real estate. Browse this site to discover lots of links and resources.

Topy Cyberspace Law Cases of 1997
http://www.gseis.ucla.edu/iclp/97cases.html

A useful site to explore to probe the emerging legal standings on a variety of contemporary issues. Developed and maintained by Professor Jerry Kang of the University of California, Los Angeles Law School. Each case features a brief, links to information, and the decision, if available online.

Copyright issues

Copyright issues have increasingly become a concern for the research administrator. Several sites offer assistance in sorting out the issues.

Copyright in an Electronic Age
http://www.landmark-project.com/copyright.html
Big Myths About Copyright Explained
http://www.templetons.com/brad/copymyths.html
US Copyright Office
http://lcweb.loc.gov/copyright/
Copyright Web Site
http://www.benedict.com
A2S2 Digital Projects
http://www.intellectualcapital.com

Task #21: Maintaining Financial Records

One of the most important administrative details when managing a research project involves financial recordkeeping. Most educational organizations have systems for recording expenditures in place. However, in the final 60 days of a project, such bureaucratic systems may not be sufficient for managing the finances and properly prevent overspending or underspending.

Tactic 21-1 Using financial recordkeeping software

Creating a financial recordkeeping system for each research project requires some commitment to details. Ensuring the accuracy of such a system allows the research administrator to effectively administer the budget.

Quicken and QuickBooks are considered the two standards in the field of personal finance and small businesses. Because these programs are frequently bundled on new computers, you may already own them. Both programs are easy to learn and use.

Quicken
 http://www.intuit.com/quicken98/
QuickBooks
 http://www.intuit.com/quickbooks/

Task #22: Selecting Mailing Systems

Despite significant advances in communication technologies, the postal system in still a necessary component of most offices. As a result, it is important to consider the many options you have for mailing and tracking your packages.

Tactic 22-1 Mailing and tracking systems

Some of the more common mail carriers are listed below. Each web site has additional information on pickup schedules, tracking packages, rates, and so on.

US Postal Service
 http://www.usps.gov/
Federal Express
 http://www.fedex.com/

UPS
   http://www.ups.com/
DHL Worldwide
   http://www.dhl.com/
Emory Worldwide
   http://www.emeryworld.com/
Airborne Express
   http://www.airborne.com/

Task #23: Purchasing Supplies

Acquiring supplies is a common activity in the administration of most research projects. Supplies may range from common (file folders, disks) to highly specialized (personalized rubber stamps, fax/modem/scanner).

Tactic 23-1 Purchasing office supplies

Office supply superstores are located in most communities.   Some of the largest companies feature web sites to assist customers in locating and ordering supplies.

Office Depot
   http://officedepot.com
OfficeMax
   http://www.officemax.com
Staples
   http://www.staples.com
Office Equipment Outlet
   http://www.oeo.com
Wholesale Supply Company
   http://www.wholesalesupply.com

Task #24 Purchasing Equipment

A research administrator working on a newly funded grant must devote some time to purchasing equipment. Many universities and large organizations will have a purchasing department, but the three tactics outlined here will assist the research administrator in preparing an initial list of options.

Tactic 24-1 Purchasing new equipment

When purchasing new equipment, it is increasingly possible to locate the vendor's home page and examine information, photos, and prices about various products. The following sites will assist you in locating a specific vendor or provide an index to potential vendors of a particular type of product.

Technology buyer's guides

The following sites provide comparative and evaluative information about technology products.

Computer Shopper NetBuyer
   http://www.zdnet.com/netbuyer/
Yahoo! Computers
   http://computers.yahoo.com/
A-List
   http://www.zdnet.com/pccomp/alist/
Maven: BusinessWeek Computer Buying guide
   http://www.maven.businessweek.com/

General equipment

The following sites provide comparative and evaluative information about technology products.

Beckly-Cardy
   http:// ww.beckleycardy.com
National School Products
   http:// www.nationalschool.com
Cyberian Outpost
   http://www.cybout.com
AVInfo
   http://www.avinfo.com

Midwest Audio Visual
   http://www.midwestvisual.com
Anthro
   http://www.anthro.com
Computer Buyer NetBuy
   http://www.netbuyer.com
Ziff-Davis MarketPlace on the Web
   http://www.zdnet.com/marketplace
Web Connection of Education Week
   http://www.edweek.org/products/products.htm
The Guide to Computer Vendors
   http://guide.sbanetweb.com
The Technology in Education Market Center
   http://www.e-biz.net/techedmc

Tactic 24-2 Purchasing upgrades

Research administrators are devoting more time to issues associated with
upgrading hardware and software, and asking questions such as, Is there a new
version of the software we use? Is it worth the effort to upgrade? If we upgrade,
will we need to add more memory to our computers? The resources in this
section will answer these and similar questions for the research adminstrator.

List of Computer Product Vendor's Internet Sites
   http://guide.sbanetweb.com
Computer Discount Warehouse
   http://www.cdw.com
TechWorks
   http://www.techwrks.com
Memory.Net
   http://www.memory.net

Power Express
http://www.powerexpress.com
   A site with information and sales on more than 6,000 batteries and portable
   accessories for mobile computers, cell phones, and camcorders.

Oil Change
http://www.cybermedia.com

A Windows product and subscription service that utilizes the internet to track current versions of software and drivers. CyberMedia maintains a master list of versions along with the latest fixes and improvements available from manufacturers. The list is then compared with the status of the versions on the subscriber's computer. When an update is identified, the benefits are explained and users can decide whether or not to upgrade their software. If you want to update, the files are downloaded and the installation is completed, but not until the existing software and configuration is backed up. A simple mouse click reverts to the previous setup in case there is a problem.

Tactic 24-3 Purchasing used equipment

The opportunity to buy and sell used computer equipment has been facilitated through the development of the web. For researchers conducting unfunded research, the need to purchase used equipment is essential. Several sites index current prices and broker exchanges between buyers and sellers.

American Computer Exchange Corp
http://www.amcoex.com
The PriceWeb
http://www.priceweb.com
Used Computer Price Index
http://www.usedcomputer.com
United Computer Exchange
http://www.uce.com

Task #25: Obtaining Technical Support

A research administrator may be called upon to obtain technical support for the research team. Three tactics are described here to assist in the process of obtaining technical support: obtaining phone support, using troubleshooting tools, and accessing web sites that offer technical support.

Tactic 25-1 Obtaining phone support

The traditional method of supporting customers is through an 800-number. To locate the appropriate number for the company you need, try the following:

American Yellow Pages
 http://www.lookup.com
NYNEX Interactive Yellow Pages
 http://www.niyp.com
The Yellow Pages
 http://theyellowpages.com
New Rider's WWW Yellow Pages
 http://www.mcp.com/newriders/wwwyp/
CommerceNet
 http://www.commerce.net

Telecommunications Research and Action Center (TRAC)
http://www.trac.org
 A web site to help consumers search for the best telephone plan. The pricing
 comparision service asks for your area code and telephone prefix, your
 average monthly long distance bill and several phone numbers you frequently call. After filling in the template, a single click brings up a chart with
 more than 20 calling plans, the cost of each call, and the total bill.

Tactic 25-2 Using troubleshooting tools

The following sites have a host of tools and resources for helping users diagnose
problems with their computer systems. Several companies specialize in troubleshooting and preventive maintenance software.

Norton Utilities
 http://www.symantec.com/
First Aid
 http://www.cybermedia.com/
RealHelp
 http://www.qdeck.com/qdeck/products/realhelp/
 indexextra.html
McAfee Total Virus Defense
 http://www.nai.com/products/antivirus/
Spring Cleaning
 http://www.aladdinsys.com

Tactic 25-3 Accessing web sites that offer technical support

Vendors are finding it more economical and efficient to create web sites that offer technical assistance, which may be in the form of FAQs (frequently asked questions) or in lists of tips.

Macintosh

Apple Support Information
    http://www.apple.com/support/
MacInTouch Resource Listings
    http://www.macintouch.com/resources.shtml
MacFixIt
    http://macfixit.pair.com/
The Mac Conflict Solution Site
    http://www.quillserv.com/www/c3/c3.html
The Mac Pruning Pages
    http://cafe.Ambrosia.com/DEF/

Windows

Healthy PC
    http://www.zdhelp.com
Windows95.com
    http://www.windows95.com
TuneUp.com
    http://tuneup.com

Other troubleshooting resources

Computer Help Desk NetForum
http://chdi.com/forum1.htm
    The NetForum is a free service devoted to answering your questions about computers or the Internet. Post your question and wait for a response from other readers or a member of the Computer Help Desk's technical staff.

Figure 7-1
The Web Site for Support Help
Used with permission of M. Victor Hanulaitis.

Support Help
http://www.supporthelp.com/
>Links to the technical support sites and home pages of 2,500 vendors. See
>Figure 7-1.

Internet Help Desk
http://w3.one.net/~alward
>Tools and advice on troubleshooting a variety of computer problems.

DukeNet Modem Users Resource Page
http://www.netcom.duke.edu/modem_user.html
>Having trouble with your modem or telecommunications software? Check
>out this Duke University site for a variety of helpful tips, installation proce-
>dures, and troubleshooting advice.

Troubleshooting Links
http://www.wellesley.edu/Computing/Binders/Dnc/links.html
>Links to a variety of helpful troubleshooting sites. By Wellesley University.

Drivers Headquarters
http://www.drivershq.com

Having trouble somewhere between the computer and a peripheral? If it is not the cable, often it is a problem with a software driver. Search this site by manufacturer, and type of product, and check for new updates.

Advice

Tip World
    http://www.tipworld.com/
Computer Related FAQs
    http://www.sparco.com/archive/cfaq.html

Task #26: Facilitating Travel Planning

Given the importance of disseminating the results of a research study, travel is often built into the budget of a research project. The research administrator may therefore need to devote some time to tasks related to travel. This section outlines web-based resources to assist in making reservations, exploring a city, checking on the weather, and working while you travel.

Tactic 26-1 Making reservations

Although a good travel agent precludes the need for these do-it-yourself tools, the research administrator can use the web to obtain a higher level of service and become a little better informed by asking different kinds of questions. Commonly, web sites are available to assist in travel planning and reservations for airlines, hotels, and rental cars.

Travel planning

Biztravel.com
    http://www.biztravel.com
Microsoft Expedia
    http://www.expedia.msn.com
Internet Travel Network
    http://www.itn.net
Preview Travel
    http://www.previewtravel.com

Travelocity
http://www.travelocity.com
TravelWeb
http://www.travelweb.com
Flifo
http://www.flifo.com
The Trip
http://www.thetrip.com
Arthur's Frommer's Outspoken Encyclopedia of Travel
http://www.frommers.com
Travel Planner
http://www.travelweb.com/hytatt.html

Reservations–General

Internet Travel Network
http://www.itn.net

Reservations–Housing

All the Hotels on the Web
http://www.all-hotels.com
Hotel World Online
http://www.hrs.de
Hotel Rooms OnLine
http://www.hotelroomsonline.com
Innstar B&B Guidebook Reviews
http://cedarcroft.com/innstar

Reservations–Ground Transportation

Amtrak
http://www.amtrak.com
BreezeNet's Guide to Airport Rental Cars
http://www.onm.com/rcar.htm
TravelScope
http://www.escope.com

Reservations–Air Travel

Best Fares
  http://www.bestfares.com
OAG Online
  http://www.oag.com
American
  http://www.amrcorp.com
Continental
  http://www.flycontinental.com
Delta
  http://www.delta-air.com
Northwest
  http://www.nwa.com
United
  http://www.ual.com
USAir
  http://www.usair.com

Travel resources

Flight Trax
http://www.weatherconcepts.com/FlyteTrax/
  This web site allows you to enter the flight number, airline, and the destination airport and it will show you the plane's route of flight, its location on
  that route, and expected time of arrival (accurate within 2 to 3 minutes) using
  FAA data. Very useful to check before leaving to pick up someone at the
  airport.

Wisconsin Medical College of Wisconsin
International Travelers Clinic
http://www.intmed.mcw.edu/travel.html
  Information for travelers planning trips abroad. Addresses pre-travel
  vaccinations, preventive medications, and health counseling.

164 Currency Converter
http://www.oanda.com/cgi-bin/ncc
  International travel? You'll need to visit this site to check on the current
  exchange rates.

Directory of Transportation Resources
http://dragon.Princeton.EDU:80/~dhb

> A massive searchable database of transportation systems around the world. View subway maps, bus schedules, and so on.

Tactic 26-2 Exploring a city

When time permits, exploring a city via the web is a marvelous tool for planning a trip. This area of the web has seen an explosion of growth and development. The resources below provide search engines for identifying your destination city, viewing maps, identifying attractions, locating restaurants, and almost anything else you would want to know about a community. Have a good trip!

Preview Travel Destination Guides
http://destinations.previewtravel.com/

> Destination guides to 87 cities around the world. Travel content by Fudor's. Extensive information, photos, and links.

Door-to-door maps

Select your starting point and your destination, and these web sites will draw a map and provide turn-by-turn directions. Useful for traveling across town or cross-country! Awesome!

MapsOnUs
> http://www.mapsonus.com/
MapQuest
> http://www.mapquest.com
MapBlast
> http://www.mapblast.com
Zip2 Yellow Pages
> http://www.zip2.com

Compass 3800
http://www.chicagomap.com

> For the truly geographically challenged, Chicago Map has developed the Compass 3800 which consists of mapping software and a hand-held Global Positioning System (GPS) receiver. Users connect the GPS to their notebook computer and then view their location on street-level maps as they travel about a city.

Mapping software

These software products allow you to create travel maps and print them to guide you on your journey when traveling by car.

AAA Map'n' Go 3.0 (DeLorme)
    http://www.delorme.com/cybermaps/cyberrouter.htm
Rand McNally TripMaker (Rand McNally)
    http:// www.randmcnally.com
Microsoft Expedia Trip Planner '98 (Microsoft)
    http://www.expedia.msn.com/
Streets USA (American Business Information)
    http://www.lookupusa.com

Food

Epicurious
    http://www.epicurious.com/epicurious/home.html
WorldWide Menu
    http://www.byte2eat.com
Eat Here: A Guide to Road Food
    http://www.eathere.com
Fudors Travel Guides
    http://www.fodors.com

Tactic 26-3 Checking on the weather

Of course, the weather is always a concern when considering travel. Many sites allow you to monitor the weather in a given locale.

Yahoo! Weather
    http://www.yahoo.com
AccuWeather
    http://www.accuweather.com
The National Weather Service
    http://www.nws.noaa.gov
WeatherLabs
    http://www.weatherlabs.com
Interactive Weather Information Network
    http://iwin.nws.noaa.gov/iwin/main.html

Figure 7-2
Mobile Computing Home Page
Used with permissionof Brian Nadel.

The Weather Channel
   http://www.weather.com
USA Today Weather
   http://www.usatody.com/weather/wfront.htm

Tactic 26-4 Working while you travel

Many researchers find themselves having to work while they travel. Although
laptop computers have made working on the road easier, it still poses special
challenges. Tools in this section will assist you in maintaining your productivity
while traveling.

The following two magazines are especially useful for helping the traveling
researcher maintain high performance:

Mobile Computing (See Figure 7-2)
   http://www.mobilecomputing.com/
mobilis
   http://www.volksware.com/mobilis

Norton Mobile Essentials
http://www.symantec.com/nme/

This product prepares your laptop for traveling by adjusting location settings and resolving connection programs. Includes a variety of useful tools to ensure that you are as productive on the road as you are at home.

Farallon's Netopia Virtual Office
http://www.farallon.com

Users set up a virtual web-based office, which they can access while traveling to use fax, message, and video conferencing services.

TeleAdapt
http://www.teleadapt.com/

The International Center on Modem Connectivity. Be sure to visit this site before you leave on your trip to find helpful information on how to stay connected with the right plugs and valuable tips.

The List
http://thelist.internet.com/

A worldwide list of Internet services and access numbers.

Hotmail
http://www.hotmail.com/

A free service from Microsoft that forwards email to a web page accessible from any web browser. Travelers can use this strategy to access their email for the cost of a local phone call anywhere in the world.

Task #27: Preparing Disaster Recovery Plans

The potential loss of research data is every researcher's nightmare. An effective research administrator develops contingency plans. A number of tools are available to assist in disaster prevention and data recovery.

Tactic 27-1 Using backup systems

A commitment to regularly backing up electronic research records is essential to ward off disaster. Although you can manually make backup copies, it is usually more reliable to purchase special software to set a timer for conducting the backups and provide media for mass storage.

Ditto Easy Drive
http://www.iomega.com

A tape backup system by Iomega simplifies the process of making backups of large hard drives.

SafeGuard Interactive Inc.
http://www.sgii.com

Remote (off-site) backup with unlimited access and storage for a fixed annual fee.

Tactic 27-2 Planning for disaster recovery

Occasionally, disaster does strike. For some reason, the data you need is not readable on the disk, hard drive, or cartridge on which it is stored. Several companies offer tools and services to restore and/or recover your data.

DriveSavers
http://www.drivesavers.com
Ontrack Data Recovery
http://www.ontrack.com
Micro Com
http://www.mcrecovery.com
Data Mechanix
http://www.datamechanix.com

CompuTrace
http://www.absolute.com

Another form of disaster is theft. An innovative safeguard against theft is the CompuTrace TRS system developed by Absolute Software, based in Vancouver, British Columbia. When installed in a portable computer, CompuTrace regularly places toll-free calls via modem to the company's office. If the owner reports the portable stolen, Absolute Software tracks down the laptop by tracing the phone call.

CyberAngel
http://www.sentryinc.com

CyberAngel from Computer Sentry protects your laptop data by using a password to lock the data. If a thief uses the modem, the system will automatically call Computer Sentry and provide caller-ID information to assist in determining the location of the computer.

Task #28: Participating in Technology Transfer Processes

Some funding sources require the research team to explore the commercial application of their work. *Technology transfer* is the generic term used to describe the process of evaluating the commercial applications of basic research findings.

Tactic 28-1 Facilitating technology transfer

Turning research and development projects into successful commerical ventures is often the responsibility of a special office within a university or business. The web sites in this section provide an introduction to technology transfer.

U.S. Patent and Trademark Office
http://www.uspto.gov

A wealth of information and resources concerning copyright, trademarks, and patents. Downloadable forms facilitate the process of applying for patents and trademarks.

Free EDGAR
http://www/freeedgar.com

A site that allows users to view corporate data filed with the Security and Exchange Commission and export the data into Excel spreadsheets. Use the search engine to create personalized lists of businesses to track.

Everything International
http://ib.philacol.edu/ib/russow.html

A directory of links to more than 100 world wide web sites that focus on global trade and business issues.

CHAPTER 8
The Quest to Become
an Electronic Scholar

Chapter 1 described the need to examine the use of technology to enhance research productivity, and provided two frameworks for transforming technology-using researchers into electronic scholars. The first framework sought to build a foundation of tools and skills that support general productivity. The second framework was based on the tasks commonly conducted by researchers.

Succeeding chapters have explored many ways in which technology can help you in your work. By now you most likely have used the worksheets provided and made some decisions about the components of your electronic toolbox. Figure 8-1 illustrates a sample toolkit for supporting general productivity when using a computer with Microsoft Windows. This toolkit includes an array of products that will support the researcher in the everyday demands of work.

Figure 8-2 is an example of one researcher's research productivity toolkit. As the researcher examined the web sites in this book, she used the bookmark feature of her web browser to save specific web addresses. Some web browsers allow the user to edit the titles or group similar sites together. The sites outlined in this figure illustrate one educational researcher's personal choices of various tools that will contribute to her work as she completes the 28 common tasks.

Your toolkit obviously will look different: it will reflect the nature of your work, your experience with technology, and the availability of specific products. Remember, the goal is not to fill up the hard drive, but rather to gain mastery over a collection of tools. The ultimate goal is to demonstrate clear and obvious gains in research productivity as a result of using technology. This will be accomplished only if we explicitly define the productivity demands we encounter and identify appropriate technology tools that can be used as part of an integrated desktop of electronic tools.

Chapter 1 provided a vision of the outcomes for technology-enhanced research productivity, but little information was provided about the journey that may be made to realize these outcomes. This final chapter provides some context for the journey that lies ahead in the quest of becoming an electronic scholar.

Figure 8-1
Sample Toolkit for Supporting General Productivity

Creating Information

Function	Tool	Product
Text	Word Processor	Microsoft Word
Graphics	Charts Paint Desktop Video	Microsoft Excel Adobe Illustrator Connectix QuickCam
Instructional Materials	Print Computer-based	Adobe Pagemaker HyperStudio; Digital Chisel

Communicating Information

In Print	Desktop Publishing	Adobe Pagemaker
Orally / Visually	Presentation Software	Microsoft PowerPoint
Electronically	Email Web Browser Video conferencing	Eudora Netscape CUSeeMe

Managing Information

In Print	Text	Microsoft Word
Electronically	Database	FileMaker Pro

Using Specialized Tools

Desktop Productivity	Planners	DayRunner
Utilities	General Purpose Virus Protection	Norton Utilities Norton AntiVirus

Figure 8-2
Sample Toolkit for Supporting Research Productivity

Professional Development

Task #1: Enhancing Knowledge and Skills
http://seamonkey.ed.asu.edu/~behrens/
http://trochim.human.cornell.edu/kb/kbhome.htm
http://www.execpc.com/~helberg/statistics.html
http://policy.rutgers.edu/cupr/SoNC.htm
http://www.alphaworks.ibm.com
http://aera.net
http://www.ed.gov/EdRes/EdAssoc.html
http://www.albany.net/allinone/
Task #2: Maintaining Current Awareness
http://scout.cs.wisc.edu/scout/
http://www.usatoday.com/
http://www.caryon.net

Designing the Study

Task #3: Engaging in Preliminary Exploration of a Topic
http://www.nwrel.org/national/
http://www.clearinghouse.net
Task #4: Conducting an Extensive Review of the Literature
http://www.alexa.com
http://www.aspensys.com/eric/
http://info.lib.uh.edu/wj/webjour.htm
http://gort.ucsd.edu:80/newjour/
http://amazon.com
http://www.carl.org/carlweb/
Task #5: Selecting an Appropriate Methodology
http://www.sagepub.com/sagepage/mtc.htm
http://www.ualberta.ca/~jrnorris/qual.html
http://www.stat.scarolina.edu/rsrch/gasp/
http://www.yforum.com/
http://www.demographics.com
http://www.nsrs.com
Task #6: Identifying Potential Sources of Funding
http://nscp.fie.com/
http://www.ed.gov/legislation/FedRegister/announcements/index.html
Task #7: Engaging in Collaboration
http://www.whowhere.com/
http://www.wpine.com/Products/CU-SeeMe/

Figure 8-2 *continued*

Task #8: Preparing Proposals
http://www.biostat.wisc.edu/grantworks/
Task #9: Participating in Human Subjects Review
http://www/psych.bangor.ac.uk/DeptPsych/Ethics/HumanResearch.html

Conducting the Study

Task #10: Creating Materials
http://www.macromedia.com/software/authorware
http://motted.hawaii.edu
Task #11: Collecting Data
http://www.palm.com
http://www.alphasmart.com
http://www.aurbach.com
Task #12: Communicating with Field Sites
http://www.eudora.com

Analyzing the Data

Task #13: Analyzing Quantitative Data
http://www.stat.ucla.edu/calculators/
http://www.spss.com/
Task #14: Analyzing Qualitative Data
http://www.qsr.com.au/
http://www.researchware.com

Reporting the Results

Task #15: Consulting Writing Resources
http://www.inkspot.com
http://askjeeves.com
http://www.uvm.edu/~ncrane/estyles
http://owl.english.purdue.edu
Task #16: Exploring the Visual Representation of Data
http://ww.inspiration.com
http://ww.inki.com/clipart
http://ww.esric.om
Task #17: Disseminating Research Findings
http://www.adobe.com/acrobat/
Task #18: Presenting Research Results
http://www.presentersuniversity.com

Figure 8-2 *continued*

Research Administrator

Task #19: Managing Information
http://www.filemaker.com
Task #20: Managing Projects
http://www.microsoft.com/project/
http://www.dayrunner.com
http://gcs.ed.gov/grntinfo/edgar.htm
http://www.landmark-project.com/copyright.html
Task #21: Maintaining Financial Records
http://www.intuit.com/quicken98/
Task #22: Selecting Mailing Systems
http://www.usps.gov
http://www.fedex.com
Task #23: Purchasing Supplies
http://officedepot.com
http://www.wholesalesupply.com
Task #24: Purchasing Equipment
http://www.zdnet.com/pccomp/alist/
http://guide.sbanetweb.com
http://www.amcoex.com
Task #25: Obtaining Technical Support
http://www.supporthelp.com/
http://www.wellesley/edu/Computing/Binders/Dnc/links.html
Task #26: Facilitating Travel Planning
http://www.biztravel.com
http://www.itn.net
http://www.weatherconcepts.com/FlyteTrax/
http://www.mapsonus.com/
http://www.epicurious.com/epicurious/home.html
Task #27: Preparing Disaster Recovery Plans
http://www.drivesavers.com
Task #28: Participating in Technology Transfer Processes
http://www.uspto.gov

Adoption of Innovations

We are in the very beginning of a predictable cycle of adoption in which the research community examines the possible benefits of utilizing technology to support its work. Indeed, as Rogers (1995) indicates, the adoption of innovation is cyclical, and the knowledge and behaviors of those involved in the adoption are influenced by a variety of factors. Moore (1991, 1995) extends Rogers' original work as he discusses five groups of potential consumers for technology products: innovators, early adopters, early majority, late majority, and laggards. Moore illustrates the adoption cycle using a bell-shaped curve that provides numeric evidence about the size of the various market segments and the demographic characteristics of each group (see Figure 8-3).

The current penetration of technology into the research curriculum appears limited to faculty considered to be innovators and early adopters. We are most likely approaching the chasm, the giant gap between the early adopters and the early majority, Moore describes that must be crossed before the mainstream routinely utilizes technology tools such as the ones we have discussed in this text.

Figure 8-3
The Revised Technology Adoption Life Cycle (Moore, 1995)
From *Inside the Tornado* by Geoffrey A. Moore.
Copyright 1995 by Geoffrey A. Moore Consulting, Inc.
Reprinted by permission of HarperCollins Publishers, Inc.

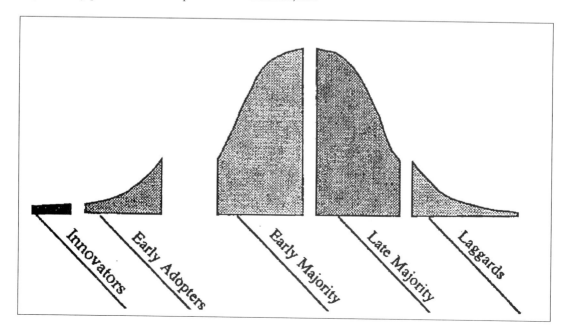

Figure 8-4
Stages of Concern Involved in Adoption of New Innovations
From: Hord, S.M., Rutherford, W.L., Huling-Austin, L., & Hall, G.E. (1987).
Taking Charge of Change. Alexandria, VA: ASCD. Used with permission of
Southwest Educational Development Lab.

	Stages of Concern	**Expressions of Concern**
IMPACT	6. Refocusing	I have some ideas about something that would work even better.
	5. Collaboration	I am concerned about relating what I am doing to what other instructors are doing.
TASK	4. Consequence	How is my use affecting my students?
	3. Management	I seem to be spending all my time in getting material ready.
SELF	2. Personal	How will using it affect me?
	1. Informational	I would like to know more about it.
	0. Awareness	I am not concerned about it (the innovation).

The process of change involves individual change, and individuals do not simply one day adopt a new technology tool and change their work habits forever. Studies point to some fairly predictable indicators of the process people will go through as they seek to incorporate new resources and tools into their repertoire. Figure 8-4 outlines the stages of concern that individuals move through (Hord, Rutherford, Huling-Austin, & Hall, 1987) as they adopt and integrate an innovation into their work. Essentially, they start at level 0 with little or no awareness of the possibilities and move through levels 1 and 2 as they gain information and decide to act on it. Level 3 involves implementation of the change and the initial management problems that arise in "making it work." Only after they have mastered the mechanics and deemed the efforts a success does their focus turn to issues of quality, sharing with others, and improvement.

This model helps us understand why workshops often meet the needs of only a small group of participants, since we fail to assess participants' stage of concern or accommodate for various levels of concerns. Regarding the use of technology to enhance research productivity, which stage best describes your immediate concerns?

The literature on innovation is important because it informs our thinking about the change process and the strategies that will be effective in using technology. The work of Rogers and Moore helps us understand the behavior of groups of people and how we need to expand professional development opportunities to meet the varied needs of different segments of our profession. The work on stages of concern by Hord and colleagues helps us understand the personal change process. Using the innovation literature will help us to move away from the individual do-it-yourself models of technology adoption and engage a larger segment of the research community in a discussion about the perceived value and use of new tools.

The Technology Integration Process

Although we now have some insight about our standing in the personal process of change, we are still without a roadmap for the journey of integrating technology into our work as researchers. The technology integration process shown in Figure 8-5 is one such roadmap.

Figure 8-5
Edyburn's Model of the Technology Integration Process

Phase 1: SELECTION	Phase 2: ACQUISITION	Phase 3: IMPLEMENTATION	Phase 4: INTEGRATION
Planning	Previewing	Organizing	Linking
Locating	Evaluating	Trainer Training	Managing
Reviewing	Purchasing	User Training	Assessing
Deciding			Extending

This model was initially developed as a result of the author's work with K-12 educators in response to the challenge of integrating technology into the curriculum. A modified version is presented here to reflect the differences between K-12 and adult education. Awareness of this model will enable researchers to properly adjust their focus as they move through the process of determining how to use technology to enhance their productivity.

The model is a recursive process involving four phases: Selection, Acquisition, Implementation, and Integration. Each phase features a number of tasks that must be completed before moving to the next phase. Whereas technology specialists typically describe their work as involving phases 1 through 3, the majority of people would prefer to focus their energies in Phase 4. The model serves to make explicit a process that is often ill-defined and provides a mechanism for various stakeholders to collaborate toward achieving a common goal– faculty and student use of appropriate technology tools in support of their personal and collaborative research agendas to improve educational practice.

Selection Phase

In the Selection phase we first seek to define our purposes for using technology (Planning) and initiate a search for appropriate technology, media, and materials (Locating). Next, we obtain reviews of the materials we have located in order to rank order our list (Reviewing). To complete Phase 1, we decide how we will act on our prioritized list of possibilities (Deciding).

In this book, Phase 1 of the Technology Integration Process has been done for you. For example, Chapter 1 presents two frameworks (see Figures 1-1 and 1-2) that define the reasons for using technology (Planning) and the types of activities in which to utilize technology. The appropriate technologies, media, and materials listed and described throughout this book have obviously already been located (Locating). Multiple reviews have been used to determine the worthiness of the technologies, media, and materials listed (Reviewing) and decisions have been made about how many resources might be reasonable to explore for each tactic (Deciding). Thus, the model partially illustrates why readers will find this book helpful in their work: rather than beginning the process at the beginning with Phase 1, as would be necessary in the current "do-it-yourself" model, readers begin the technology integration process in Phase 2.

Acquisition Phase

The goal of Phase 2, Acquisition, is to review products with the intent of purchasing them for immediate use. The researcher begins by determining how to view and try out the various technologies, media, and materials (Previewing). This task is achieved very easily through the web as a result of preview centers and downloading of demo copies, and becomes a little more difficult when the need arises to locate a sales rep and schedule a visit. The researcher examines each product and conducts a formative evaluation to determine the product's ability to meet his or her needs (Evaluating). It is during this stage–the continual previewing and evaluation of product after product–that the researcher may feel bogged down. When a suitable product is identified, the researcher moves on to the next task, which involves the administrative procedures of ordering a product (Purchasing).

Implementation Phase

Phase 3 is a significant milestone: the researcher now owns the product. The first task in the Implementation Phase, Organizing, involves deciding where to store a software product (e.g., individual machine, designated station, file server, etc.) and the administrative tasks of taking inventory of and supporting a product. Next, the various individuals that will teach others to use a product should acquire the mechanical skills to operate and navigate a product, learn basic troubleshooting skills, and explore curricular applications (Trainer Training). In the do-it-yourself model, this means, Teach yourself! When working in an environment where no one else is interested in using technology, the researcher needs to teach him- or herself by using the manual, doing the tutorial, finding a friend who can help, and so on. Finally, the researcher needs to make provisions to train the end users (e.g., students, members of the research team) how to access the product and navigate the various components successfully (User Training).

Integration Phase

Notice that until now the researcher has spent most of the time completing the tasks involved in Phases 1, 2, and 3 and little or no time involved in actual research. It is no great surprise, then, that many researchers get lost along the way, or, claim that "it is too much trouble." Indeed, it is not until Phase 4, Integration, that the researcher reaps the dividends of the investment made during the first three phases. It is in this phase that the researcher turns away from a preoccupation with "stuff" and refocuses on the productivity or research tasks. The researcher must decide when a product will be used (Linking) and select an appropriate management plan (Managing) on how users will access and use the technologies (e.g., laptop computers, designated workstation, network, dial-in access, etc.). The researcher needs to evaluate the success of the experience and make corresponding changes if necessary (Assessing). Finally, unless ways are found for creatively re-using the technologies to allow the researcher to stay in Phase 4 (Extending), the process begins all over again by identifying a new objective (Planning).

Given the significant commitment of time and energy involved in this process, it is easy to see why so many people get discouraged along the way. Unfortunately, technology integration is still a do-it-yourself project. Hopefully, this book will expedite your travel through Phases 1 to 3 so that you can devote your time and energy to your research in Phase 4. Use this roadmap to determine your progress on your journey toward electronic scholarship and to focus your attention on the important benefits that technology will provide your work once it becomes integrated into your daily routine.

Looking Ahead

As discussed at the end of Chapter 1, several activities remain before readers can become fully certified "Electronic Scholars." These activities fall under the final goal–Goal 3: Rethink how educational researchers work in light of available information technologies.

Strategy 3.1 Facilitate training in the use of the toolbox

In order to develop one's full potential as an electronic scholar, each researcher must become skilled in using the tools selected for his or her toolbox. Most of us use only a fraction of the features available in a software program. Make a commitment to go beyond the initial training on the mechanics of a program. Periodically, participate in training that will allow you to take advantage of intermediate and advanced features of the various tools. Your aspiration should be to exploit the power of each tool in your toolbox.

Strategy 3.2 Foster discussions of advanced strategies for using the toolbox to increase researchers' effectiveness and efficiency

Related to the issue of training is the need to foster discussion of advanced strategies for using the toolbox to increase researchers' effectiveness and efficiency. This book outlines a lot of different avenues to take. We need much more discussion and many more examples of how individuals work differently as a result of technology. Although Fetterman (1998) provides a very powerful personal statement about the ways he has found technology enhances his work as a researcher, we need to know much more about the productivity strategies associated with these new tools. Given new ways of working, how do we know they are more effective? We also must be on guard against the explosion of specialized tools that spread our attention among such a large collection of tools that we fail to master a core set of powerful but flexible tools.

Strategy 3.3 Validate the toolboxes and productivity strategies in order to reduce the do-it-yourself nature of the technology integration process

The challenge of moving beyond technology integration as a do-it-yourself project will require the research community to move beyond individual visions to common visions. If we can agree that the technology integration process outlined here (see Figure 8-5) defines the essential tasks, we can then begin to develop collaborative strategies for helping others move through the process. Consider how the following examples might facilitate the process:
   • A newsletter publishes an article comparing and constrasting several qualitative research products. (Reviewing)

- A university negotiates a site license for a new software tool for a group of researchers. (Purchasing)
- During a conference presentation, a researcher shares examples of protocols that were used to help participants remember key commands while they engaged in an interactive online data collection activity. (User Training)
- A research team discusses how laptop computers will be used in the field and who will have responsibility for uploading the data and making backups. (Managing)
- An electronic discussion group exchanges tips for using a research tool. (Managing)

As stated earlier, there is an urgent need to move beyond individual visions (such as those described in this book) to common visions. We need to foster a dialogue about the essential nature of technology to research productivity with toolboxes that have been validated by the research community.

Prior to the publication of this book, researchers interested in using technology to enhance their work would begin the integration process at the beginning of Phase 1 (Planning) and engage in the tasks as a "do-it-yourself" project. The frameworks and resources organized in this book allow readers to begin the process at Phase 2 (Previewing). However, the development and dissemination of validated toolboxes would allow users to begin the process near the end of Phase 3 (User Training). This strategy represents a considerable savings of time and energy for the research community. And, if these efforts are to be any more than a shopping exercise, we must focus not only on the tools but on the Phase 4 productivity strategies that provide demonstrated valuable outcomes for our efforts.

This book represents a toolkit of possibilities; individual researchers and the profession must take up the challenge to define an essential toolkit. I hope you will join others in the research community in the quest to collectively and with conviction provide answers to the question, "How does technology enhance research productivity?"

References

Fetterman, D.M. (1998). Webs of meaning: Computer and Internet resources for educational research and instruction. *Educational Researcher*, April, 22-30.

Hord, S.M., Rutherford, W.L., Huling-Austin, L., & Hall, G.E. (1987). *Taking charge of change.* Alexandria, VA: Association for Supervision and Curriculum Development.

Moore, G.A. (1991). *Crossing the chasm.* New York: HarperBusiness.

Moore, G.A. (1995). *Inside the tornado.* New York: HarperBusiness

Rogers, E.M. (1995). *Diffusion of innovations* (4th ed.). New York: Free Press.

Index

A

administration of research projects, 141-164
adoption of innovations, 170-171
analyzing data, 107-115
assessment instruments, 78
audio and video broadcasts, 50-51

B

backing up files, 44-45, 162-163
bar coding software, 101
book publishers, 64-66
browsing, 37-38, 55-60

C

career opportunities, 35-37
charting and graphing software, 124-125
clip art, 126-127
collaboration, 82-88
collecting data, 96-105
college and university home pages, 36
communicating with field sites, 105
compliance with technical requirements,
    146-147
computer-based data collection tools, 99-101
copyright, 120, 147
creating materials
    CDROMs, 138-139
    print materials, 92
    rapid prototypes, 93-94
    web-based materials, 95
current awareness, 45-52

D

data analysis, 107-115
data bases
    bibliographic, 44
    creating, 44, 144
    searching, 60-62
data collection, 96-105
data collection probes, 99-100
data sets, 26
daily lessons, 40
desktop publishing, 129-130

desktop scanner, 142
disaster recovery, 162-163
digital cameras, 104-105
dissemination, 32-33, 129-139
document delivery services, 64-67

E

electronic
    dissemination, 137-139
    forms, 144
    information services, 45-47
    mail, 54, 105
    portfolios, 101, 137
    scholar
        becoming an, 165-176
        creating toolkits, 8-9, 166-169
        defined, 3
        integrated desktop, 5
        productivity strategies, 175
    text, 5, 43, 142-143
    textbooks, 24, 31, 70
ergonomics, 52
exploring a city, 159-160

F

fax dissemination, 136-137
fellowships, 36
file compression, 143
filing, 43
financial recordkeeping, 148
food, 160
frequently asked questions, 41
funding alert services, 79-80

G

general productivity, 6, 13-14, 166
grammar, 123-124
grant
    funding, 79-82
    proposal development, 88-89
graphing calculators, 108

H
hand-held input systems, 96-97
human subjects review, 89-90

I
institutional research boards, 89-90

L
language translation, 122
learning about research, 22
listservs, 45-47
locating
    an individual, 82-83
    factual information, 118-120
    phone numbers, 153

M
mailing, 148-149
managing
    information, 142-144
    financial records, 148
    projects, 144-145
mapping software 128-129, 159-160
modeling software, 128-129
monitoring changes on the web, 51-52, 134

N
news services, 47-51

O
offline browsing, 52
online classes, 41-43

P
paint programs, 125-126
participating in the research community,
    33-35
personal digital assistants, 96-97
personal library, 43-45
personalized news services, 49-50
planning a research study, 53-90
portable keyboards, 98-99
position announcements, 35
preliminary exploration of a topic, 54-61
presenting research results, 140
professional associations, 54

professional development, 21-52
project management, 144-145
purchasing
    equipment, 150-152
    supplies, 149

Q
qualitative data analysis, 112-115
quantitative data analysis, 108-112

R
rapid prototype tools, 93-94
ready reference, 118-120
relocating, 37
reporting the results of research, 117-140
research administrator, 141-164
research design, 68-69
research methods
    and procedures, 26-33, 69-78
    education, 28, 33
    geology, 27
    history, 28-29
    interdisciplinary, 29
    life sciences, 27, 32
    qualitative, 69-71
    quantitative, 72
    social sciences, 30
    survey, 73-75
research process
    designing the study, 53-90
    conducting the study, 91-105
    analyzing the data, 107-115
    reporting the results, 117-140
research productivity, 7, 10-13, 15-17, 167-169
researchers knowledge and skills, 22-52
reservations, 156-158
review of the literature, 54-68
revised technology adoption life cycle, 170

S
searching with
    directories, 37-38
    search engines, 38
    metasearch sites, 39-40
self-directed learning, 37-43

software
    bar coding, 101
    bibliographic databases, 44
    charting and graphing, 124-125
    clip art, 126-127
    computer-based data collection tools,
        99-101
    data bases, 144
    desktop publishing, 129
    electronic forms, 144
    electronic portfolios, 101, 137
    file compression, 143
    geographic information systems, 128-129
    mapping, 128-129
    modeling, 128-129
    paint programs, 125-126
    project management, 144-145
    qualitative data analysis, 113-115
    quantitative data analysis, 108-112
    rapid prototype tools, 93-94
    text conversion, 142-143
    time management, 145
    web site, 135-136
    work group, 87-88
    unobtrusive computer-based data
        collection, 102
    video capture, 139
stages of concern, 171-172
statistical calculators, 108
style guides, 121-124
subject selection, 76-78
survey research methods, 73-76

T
teaching about research, 23-26
technical support, 152-156
technology buyer's guides, 150
technology enhanced productivity
    general productivity, 6, 13, 166
    goals and strategies, 3-18
    integrating tools into daily work,
        174-176
    research productivity, 7, 13, 167-169
    rethinking, 2-4
    taxonomy of technology enhanced
        research productivity, 10-12

technology integration process, 172-174
technology toolboxes
    in business, 1
    in education, 2
    general productivity, 14, 166
    research productivity, 15-17, 167-169
technology transfer, 32, 164
telephone billing plans, 153
text conversion, 142-143
time management, 145
travel
    planning, 156-162
    reservations, 156-158
    working while traveling, 161-162
troubleshooting tools, 153-156

U
upgrading hardware and software, 151-152

V
video capture, 139
video conferencing, 84-86
video training, 43
visual representation of data, 124-129
voice input, 103-104

W
weather, 160
web-based
    data collection, 102
    publications, 62-63
web demographics, 134
web publishing, 131-136
web sites
    creating, 132-135
    maintaining, 135-136
web telephony, 83-84
webcams, 31, 85
word usage, 122-123
worksheets
    general productivity, 14
    research productivity, 15-17
    world wide web self-assessment, 19
writing resources, 118-124